Practical Instructions for Millers and Carpenters for Constructing Mills

Prepared by Heinrich Ernst, Master Millwright

Volume 2

With 17 copperplate engravings

Leipzig

Gerhard Fleischer, Jr.

1805

Translation by Karin I. Knisely

Technical Annotations by Thomas P. Rich

Contents

Chapter	Topic	Page
Foreword	By Dr. Melzer	2
Chapter 1	Leveling and Calculations of the Water Quantity Delivered by a River or Stream over a Certain Time	4
Chapter 2	Building a Staber Mill: Design and Practical Experience	15
Chapter 3	Building a Staber Mill with 4 Millstones	36
Chapter 4	Sluices	45
Chapter 5	Weirs	55
Chapter 6	Panster Mills	61
Chapter 7	Building a Strauber Mill with Two Millstones	78
Chapter 8	Proper Handling and Evaluation of Wood for Mill Construction Description of an English Tannin Removal Machine	84

Foreword by Dr. Melzer

When someone other than the author of a book writes a foreword, then it must pertain to something that the author was not able to say properly himself. This is also the case here.

The writer of this book, of which this is the second volume, is indeed a man who knows what he's talking about on this subject. Not only is he familiar with the necessary mathematics and other associated sciences, he also knows enough about what is practical and doable, since he was originally a skilled miller and millwright.

In his art and science, however, the knowledge of what is applicable is more essential than in mechanical engineering, under which mill construction also falls. Even if the admirable formulas of a Langsdorf, Mönch, and other meritorious men are correct, experience from these men and all their blind copycats has taught that the construction did not always meet expectations. Well, how could it be otherwise? Nature is not alone infinitely diverse in the creation of physical and spiritual objects, rather also just as diverse in the pronouncement and backlash of forces. Here the algebraic formula does not yet suffice; rather experience with its amendments and changes is also required.

How many have hurt themselves in general by rushing into construction! How much hastiness doesn't occur every day, especially in mill construction? What publication and handbook about mills unites the knowledge of our ancestors and the experiences and corrections of today's contemporaries? There is still a need for this kind of textbook. The writer of this book is as willing as capable to meet this need.

But what a sad fate the author was dealt and is still dealing with! Nothing is more valuable than to carry out experiments, and the rich man is exactly the least inclined to experiment, since he is not tempted by a desire to improve.

Furthermore, when an experiment fails, there are hundreds of judges, who, with highbrow expression, now know that it won't work like this. But not a single one of them was clever enough before to be able to correctly predict the outcome of an experiment. Ignoble enough are then the many of these kinds of people who call these men fools, who dare to do experiments for the betterment of mankind. And so where are the extolled powerful and rich and economic and naturalist organizations, which authorized experiments as overhead costs? I think their experiments were not of any significance. No one can accuse me of misanthropy or rudeness when I am only telling the truth! I also contributed according to my means so that the writer of this book, after almost ending up poverty-stricken as a result of his education and experiments, was able to increase and expand his knowledge to the current state.

But, since this support only benefits princes, ministers, and the rich, then I call upon them to relieve me! I cannot repay this man for his knowledge and noble character. Just like an unpolished jewel, he possesses little luster, but inner worth to understand, guide and provide instructions for large machines; he combines theory and practice.

Currently, he is working on understandable instructions and remarks for mechanics, craftsmen, and millers to calculate mechanical power. No detailed discussion, regardless of its quality, would be necessary if the mechanisms were more widely known. However, the presentation and the formula must be suited to the understanding of the craftsman. In addition, he is going to publish a special machine and mill newspaper. For this purpose, he is going to talk with various academics and experts.

At no time like the present has the need and value of machines been recognized. It would be shear folly for a traveler to take the long and difficult route instead of the shortest and easiest one. It would be just as foolish not to want to reduce and ease as much as possible our difficult work! Of course, the shortest route is also the most difficult to find.

Man was created to exercise his mental faculties as soon as his animal instincts have been satisfied. Oh, if only I could relieve mankind of all oppressive work, particularly those who heave a sigh in the raw climate of the North and South Pole! When man stops being a beast of burden, then his complete awe of the Creator and Nature will be renewed; then man will become friendlier and gentler! This is because oppressive hardship and work make him rough and uncharitable and produces resistance among the dimwitted rich.

Every machine that alleviates and shortens toil benefits mankind. Anyone who doubts this would have to curse and eradicate henceforth all machines. Plows, farming utensils, factories, and mills are nothing but machines. If, based on experience, they already were useful in the imperfect form of mankind, oh brothers of princes, ministers, and rich men of every social level – do as much as is possible!

Money in the coffer is dead and without spirit; but putting it to good use has an effect into eternity! Who can determine, what little bit and what amount of money bears the most fruit!

Adolph Heinrich Melzer, Ph.D.

Chapter 1

Leveling and Calculations of the Water Quantity Delivered by a River or Stream over a Certain Time

Part 1

The fundamental things that a practitioner needs to know about building a watermill require knowledge of water's efficacy, because this is the cause of the motion of such mills.

The things that the practitioner must understand completely and be able to apply are as follows:
1) Leveling
2) Determining the water flow per unit time
3) Positioning the water wheel at the correct height, depending on local circumstances, taking the fall into consideration
4) Knowing how to correctly determine the gear ratio based on the height of the water wheels.

If the practitioner understands these things, then the mill will not fail to function as it should, providing he applies them correctly.

Part 2

It is an established truth that the power of the water that turns the water wheel comes from the fall (or slope) of the water flowing in a river. What do we actually mean by fall? Imagine two straight lines a d and c d (Plate I, Fig. 1) from the center of the earth to its surface a b. Then let the line a d or c d move about the center d. In this way, the line's endpoint a or b describes the arc a b and the lines a d and d b are the radii of this arc a b or the largest radius of the earth. This arc is now called the true horizontal line. Now imagine another straight line d f[1], which is tangent to the arc a b (or the true horizontal line) at only one point g; this line would then be called the astronomical or apparent horizontal line. Furthermore, imagine a river originating from a, whose free surface a touches the true horizontal line and whose discharge flows downward over a certain distance, e.g., 5 or 600 ells[2] downstream, and

---

[1] Translator's note: It looks like the line is e f in Fig. 1. Technical Annotation: Correct
[2] Translator's note: One ell is the distance between the elbow and fingertip. In the North, often 2 feet; in Prussia 17/8 feet; in the South variable, often 2½ feet. Source: http://en.wikipedia.org/wiki/German_units_of_measurement#Elle_.28ell.29

Technical annotation: Early European measurements were often related to the lengths of parts of human arms. The el (or ell) has been referenced by the translator as the distance from the fingertip to the elbow. The translator's note refers to values used in various regions in 18th and 19th century Germany. From Ernst's area examples in this chapter, his value for the el was about 19 inches in length. Other early references give the el as twice the distance from the fingertip to the elbow or from the fingertip to the arm pit. This would account for

whose surface is c b lower than that at the true horizontal line a b at a. This elevation change c b is the fall over the given length of the river. Because water cannot flow without a slope, a river without fall would be a still body of water and unusable for moving machinery.

Part 3

The fall of a river is determined with a geometric measurement called leveling or spirit leveling. Leveling is a very precise measurement and must thus not just be started with the greatest precision, but also completed with the same precision. If this is not done, errors will occur which can make the goal difficult.

Part 4

The instruments used to determine a horizontal line are called leveling instruments. Before one uses these instruments for leveling, however, one must be convinced of their accuracy in order to ensure that the horizontal line determined from every given point makes a right angle with the direction of gravity. Once you are certain about this, then the horizontal line determined with such an instrument will also be accurate.

   1) Note

The apparent and the true horizontal lines can be found using leveling. §. 2. The apparent horizontal line alone is not useful when building a mill. It is only used in situations when leveling over long distances of approx. one or half a mile and the fall is not as critical as when building a mill, or in disputes over mills and water rights, where a lost inch of fall costs a lot of money. For these reasons, the true horizontal line is more important than the apparent horizontal line and that is why I am going to show you how to determine this line for the aforementioned purpose as precisely as possible with inexpensive instruments.

Part 5

The least expensive instrument most often used by millers to measure fall is the so-called plumb-bob level (Schrot- or Setzwage). Leveling with a plumb bob is easy and many millers are quite familiar with it from using Beyer's and Melzer's mill construction books, in which instructions for using this instrument are described in detail. For this reason, it is unnecessary to explain its use here. There is just one thing that I have to comment on, and that is, how to be certain of the accuracy of a plumb-bob level and how it should be equipped to achieve its purpose.

Part 6

As is known, the plumb-bob level consists of a precisely planed rectangular board (Richtscheid) a b, Plate I, Figure 2. This board can be 1 ell, 14 to 16 inches long, and 4 to 5 inches in Dresden units of measurement (Dresden Maas) wide. Then a tongue c d is mortised cleanly at a right angle into the

---

larger and varied average values in inches given for other European countries: England ≈45, France≈54, Poland ≈31, Prussia ≈25.5, and Scotland ≈37. Note: In later chapters, mill dimensions indicate that Ernst's ell = 24 inches there.

center of this board. This tongue can be 1 ell, 6 to 8 inches long from the upper edge of the board a b, and 3 to 4 inches wide.

So that this tongue remains tight and doesn't move, and is not easily subjected to warp, two jaws e e are attached in a groove in the angle formed by the tongue and the board. Then a small ivory plate is positioned in the middle of the tongue e d[3] and a vertical line is drawn onto it. Finally, strips f f with grooves are glued onto both sides of the tongue. A glass plate is inserted into the grooves so that the pendulum is not disturbed by the wind and it can hang even with the vertical line during use.

2) Note

A plumb-bob level that is used for leveling must be constructed of a good quality, dry wood, e.g. white or red beech. Oak can also be used. Even if this kind of wood seems to be completely dry, it's still better to put it in a not-too-warm oven so that the level doesn't warp. The wood used for this level and which shouldn't warp becomes even better when the tannin is removed. This process is so important that I feel compelled to describe at the end of this book a very useful tannin-removal machine, which I have in my own workshop. Wood from which tannin is removed with this machine does not warp and is highly usable for building machines and mills, in particular the wheels.

Part 7

The previous section described the parts of a plumb-bob level and how it must be constructed for leveling. The next point is how one can be sure that it is accurate; in other words, the vertical line always forms a right angle with the horizontal line whenever the level is used. To convince yourself, place the level on a level surface, such as a table, and put two small blocks under the ends of the board a b in Figure 2. These blocks must be positioned in such a way that the plumb line lies exactly on the vertical line when the two ends a b sit on the blocks. When this is the case, the level is horizontal. Now turn the level around so that the b-end now stands where the a-end was previously. Make sure, however, that the blocks are not moved at all.

If the plumb line again lies exactly on the vertical line when the level is in this position, then this is a sign that the level is accurate.

Part 8

Another readily available instrument for leveling is the protractor, which is used by mine surveyors. This protractor consists of an arc a b d, Plate I, Figure 3, made of lightweight brass. A small weight d hangs from a thin string from the midpoint c. The two quadrants formed by the periphery c d a and c d b starting from the midpoint c are each divided into 90 degrees. There are two hooks e e, which are bent in opposite directions, along the diameter a b. The hooks have a gap that allows the protractor to be attached to a string with clothes pins when the string is not perfectly horizontal. The following paragraph describes how the protractor is used for leveling.

---

[3] Translator's note: In the previous paragraph, the tongue was designated c d. Technical Annotation: Correct

1) Hammer a post *a* into the ground at the site where you want to begin leveling. Screw an iron wood screw with an eye, shown in Fig. 5, into the post at a point deemed appropriate for the task at hand. Attach one end of a string to the eye, and turn the eye so that it lies in a horizontal plane. Attach the other end of the string to another such wood screw, which has been screwed into a second post *b*, which has been hammered tightly into the ground, so that when the protractor *d* is hung on the middle of the string, the plumb line hangs precisely on the 90 degree mark. Under these circumstances, the two points e and f, where the screws are located, are on a horizontal plane. Measure the height from e to the free surface of the river c and write this value down as "upstream;" now you are finished with the first station.

2) Leave post *b* in the ground and move the first e^4 to another position, given by *g* in Fig. 4, determined by the length of the string. Hammer post *g* into the ground, attach the string to the screw eye on post *g*, and suspend the protractor from the string. Tighten the string and raise or lower it until the plumb line lies on the 90 degree mark. In this position, points *g* and *f* lie in a horizontal plane. Follow these steps until the desired distance has been leveled. If, however, it is not possible to get a continuous horizontal line from the first leveling point, due to unevenness of the terrain, then one follows the procedure for the plumb-level bob. One climbs or descends[5], as required by the circumstances, and writes it down. In the end, when you subtract one sum from the others, the difference represents the fall height.

3) Note.
Leveling with the protractor is very effective and easy, but only when wind does not affect the lightweight plumb line. The plumb-line bob is preferred when it's windy.

Part 9

When there are disputes involving mills, there are often situations in which the fall must be determined within a quarter of an inch, as I know from experience. Furthermore, the disputed stretches often are only 6 to 800 ells long. In these cases, the following method is recommended. We assume that the water level in a river or stream will stay constant for 15 minutes. Hammer a post into the ground every 100 ells or so and level each stretch. In this manner, you can determine the fall not just over one stretch, but quite accurately over the entire distance.

---

[4] Translator's note: I think the author means *a* (the post) rather than *e* (the screw eye) Technical Annotation: Correct

[5] Technical annotation: The expression "climbs or descends" refers to moving up or down the stream and adjusting the height of the leveling string on successive posts. For example in Fig. 4, as the process of setting the leveling string moved down stream to the right, the height of the string is shown lower below post g than above to make the required measurements easier. Note that the absolute height of the leveling string is not important because the fall between two successive posts is simply the distance from the string to the top of the water as measured at the down-stream post minus the distance as measured at the up-stream post. The values of fall as measured at the post intervals can then be added to give the total fall between the very first and last posts in the sequence. When divided by the length between these two extreme posts, the result is the fall per length of stream in the vicinity of the measurements (ex: inches of fall per foot of stream).

Part 10

When building a mill, it's not enough just to know how to measure the fall of the river where the mill is to be built. You must know how to determine how much water the river will supply in a given time to decide what kind of water wheel is appropriate for that amount of water. A measurement called "Consumptions" or velocity measurement is used for this purpose. Someone who is unfamiliar with this concept will probably ask: What kind of measurement is that? The answer is: One that lets you determine how much water flows in a minute or a second knowing the velocity of the water and the height of the vertical cross section of a river. Once the water quantity is known, then, given enough experience, it is easy to calculate how many cubic feet of water per minute or second are required for one or another type of water wheel to make the mill work well. This knowledge can then be applied to the measured fall, which will be explained clearly in the next section.

Part 11

To calculate the water quantity per minute or second, the cross section of the river must be determined. As I mentioned previously, someone who is unfamiliar with geometry will find this difficult, so I will try to give an understandable explanation. When measuring a river profile, the main thing is to measure the width of the river and the various water depths correctly, and then to convert the data into the desired units of area. For the measurement, choose a site where the width of the river is the same, as much as possible, over a distance of 2 to 300 ells, and where the flow is about the same. At these sites, one will find that neither sand banks nor other uneven surfaces can be observed, so that the water flows on a pure gravel bed. At these kinds of sites, the water velocity is fairly even across the depth as well as on the surface. There is an instrument called the Pitot tube. This instrument can be used to measure the velocity of the water in the middle and also near the bottom. This instrument requires knowledge of geometry and nature and, if one heeds the previously described local conditions carefully while measuring velocity, is not really necessary. Because most of my readers would not be able to understand how this instrument works from my description, I refer those of my readers who are familiar with geometry and nature to Belidor's *Architectura hydraulica*, Vol. 1, Art. 614 or to Huth's *Construction of and Guide to Watermills* Part 10 and other similar works, in which this instrument is described and reproduced. In order to stay on topic, let's turn our attention to an easier way to measure the velocity as it applies to our purpose.

Part 12

Here is the method used to measure the discharge of a river:

Stretch a string a b (Plate II, Fig. 1) across a river so that it is parallel to the surface of the water and at right angles to the bank, and attach the two ends to posts c and d. This string must be such that it is precisely marked with feet and inches using small wire rings, as shown in Fig. 1. Now, sitting in a small boat, measure the distance e f in Fig. 1 using a measuring stick that is also divided into feet and inches and is weighted down with lead. Immerse this measuring stick vertically until it rests on the bottom of the river and measure the depth to the surface of the water e f.[6] Using the scale bar, this depth corresponds to 3 feet, 11 inches.

Now continue measuring the depth either towards bank b or bank a. Also measure the distance from posts c and d to the location e where the first depth measurement was taken. Write down this distance, as well as all subsequent distances and depths, including how far they are from each other, on a piece of paper. For example, in our diagram, the distance from the bank a at the surface of the water to e is 6 feet, 3 inches. After noting the point of the first depth measurement, proceed toward one or the other bank at intervals of 3 to 5 feet, lower the measuring stick into the water until it just touches the bottom, and note whether the depth, for example, at g is the same as at the first location, or if the river bed is higher or lower. Assuming that the river bed is uneven, then the depth g h at a distance e g from the initial measurement point would have to be measured and noted. Repeat this procedure when the river bottom is uneven, as exemplified by the depths l m, i k, e f, and g h in Fig. 1.

After the water depths and their corresponding horizontal distances have been measured and written down, we can make a cross section of the river as follows.

Part 13

Put a sheet of paper on a drawing board and include an appropriate scale bar. Then draw a straight line a b (Fig. 2), which represents the surface of the water. On this line, draw the width of the river to scale, as in a and b in Fig. 2, corresponding to the points at which the water touches the two banks at a and b in Fig. 1. Again to scale, mark the distance from point c on the bank[7] to the point e where the first depth measurement was taken and add a line for water depth e f. In the same manner, add the water depths at the other intervals. When the points are connected as in the figure, then one has a scaled-down version of the river profile, as can be seen clearly in Fig. 2.

Part 14

After drawing the river profile according to the aforementioned procedure, the cross-sectional area is calculated as follows. From geometry it is known that the product of half of the base of a triangle and the height is the area. Alternatively, the area of a triangle can be found by multiplying the base by the height and dividing the resulting product by 2. The other shapes formed by the parallel depth lines are transformed either into true rectangles or squares by adding a line down the middle (*Linea intermedia*). A shape that is enclosed by four straight lines, of which only two are parallel, and the other two have different lengths and form different angles, is called a trapezoid. The lines n o, p q, and r s shown in our river profile in Fig. 1 and 2 represent the lines added to allow the cross-sectional area to be calculated. To calculate the area of a trapezoid, multiply the length of the line n o in Fig. 1 by the width l i. A sample calculation is shown in the next section.

---

[6] Translator's note: The distance e f in Fig. 1 is not the same as the water depth w f. Technical Annotation: Correct

[7] Translator's note: In Part 12, *c* represents the post and *a* the point on the bank. Technical Annotation: Correct

Part 15

In our river profile, we need to calculate the area of 2 triangles and 3 trapezoids, based on 12 inches in a foot. The steps are:

For triangle t m u.	Base t m =	35	inches
	Half height t u =	9	
	--		
	Area of triangle	315	square inches
For trapezoid t,v,m,k.	Line n, o =	40	inches
	Width t, v =	25	"
	--		
		200	
		80	
	--		
	Area of trapezoid	1000	square "
For trapezoid v,w,k,h[8]	Line p, q =	46	inches
	Width v, w =	31	"
	--		
		46	
		138	
	--		
	Area of trapezoid	1426	square "
For trapezoid w,x,f,h.	Line r, s =	46	"
	Width w, x =	40	
	--		
	Area of trapezoid	1840	square "
For triangle x,y,h.	Base x h =	46	"
	Half height x y =	22	"
	--		
		92	
		92	
	--		
	Area of triangle	1012	square inches

---

[8] Translator's note: I think this should be v,w,k,f. Technical Annotation: Correct

Sum		
Triangle t,m,u	315	square inches
Trapezoid l,v,m,k[9]	1000	"
Trapezoid v,w,k,f	1426	"
Trapezoid w,x,f,h	1840	
Triangle x,y,h	1012	
Total cross-sectional area of river	5593	square inches

[Calculation shown on p. 20 of German original]
To convert square inches into square feet, divide by 144, which gives 38 square feet, 121 inches.[10]

Part 16

Now, it's not just enough to know how to measure the cross-sectional area of a river; one must also know how to calculate how much water flows through this area in one minute or one second. To do so, one measures the velocity of the flowing water. We know from experience that the water velocity in a river is seldom the same at depth as on the surface. Therefore, when we need to determine the discharge precisely, we use the Pitot tube, described in Part 11, for which the user must have the appropriate knowledge and training. However, because when designing a mill, a few cubic feet of water don't make that much difference, we will use a procedure described by the famous Mr. Busch to determine water velocity. His procedure uses a less expensive instrument, which I will describe and then show how it is used.

Part 17

This instrument consists of a circular disk a b (Plate III) made of linden wood, 12 inches in diameter and ¾ inch thick. A small hole is drilled through the center of this disk and a thick wire c is inserted into the hole. A cylinder turned from hardwood and which is filled/coated (ausgegossen??) with approx. one pound of lead, is attached to the bottom of the wire so that the instrument is held in constant tension and is not disturbed by the wind. In order to be able to raise and lower the cylinder depending on the water depth, the hole in disk a b is made a little larger than the width of the wire and a small wedge is driven into it. The wedge allows the cylinder to be adjusted to the depth of the water. However, the wire a b must be perpendicular as much as possible to the surface of the disk[11]. In order for this to happen, the wedge must not be driven tighter than necessary so that the wire can still move a little.

Part 18

Here is the procedure for using this instrument: Go to the site where the string was stretched across the river and the depths were measured. Get a clock or watch with a second hand, or, if one is not available, a seconds pendulum that beats every half-second. This seconds pendulum was described in detail in Vol.

---

[9] Translator's note: I think this should be t,v,m,k. Technical annotation: Correct, and

[10] Technical annotation: should be square inches. (Often in older literature authors neglect to show the square on area units.)

[11] Translator's note: It seems more logical for the "a b" to follow the word "disk" than the word "wire."

1 of my *Practical Mill Construction* book in Chap. 8, Part 6. Now get into the boat with the instrument, paddle upstream a little ways, and put the instrument in the middle of the river, allowing it to be carried by the water current. At the same time, a second person standing at either post a or b (Plate II, Fig. 1), where the river profile was measured, must be prepared to watch when the top tip of the wire crosses the line a b between the two posts. At this exact moment, the seconds pendulum is started, assuming that it can also be stopped (or note the position of the second hand on the watch). Count the number of swings (or note how many seconds have passed). A third person holding a rod walks along the bank, following the swimming instrument, so that the line from the instrument to the person is as perpendicular as possible to the line of the bank. After an arbitrarily chosen number of seconds, the second person (who is watching the time) signals the third person, who puts the rod into the ground at that exact moment. The distance from this point upstream to post a or b is measured and the water velocity per second is calculated using the rule of proportions, as described in the following section.

2) Note

For a more accurate determination, it's better to measure velocity twice and then take the average. For example, let's say the water velocity was 30 feet per minute in the first trial but only 26 feet per minute in the second. Add these two numbers together and halve the sum to get 28 feet per minute as the average water velocity.

Part 19

Let's assume that, using the aforementioned instrument, the water velocity in 30 seconds or half a minute was 60 feet. Then the velocity in one second would be

30 sec / 60 feet = 1 sec / x

[Calculation on p. 24 of German original]   x = 2 feet

One multiplies this velocity of 2 feet per second by the cross-sectional area of the river to determine the water quantity that flows through this area per second. For example, as determined in Part 14, the cross-sectional area of our river is 5593 square inches. This figure multiplied by 2 feet or 24 inches gives 134,232 cubic inches. This figure divided by 1728 (the number of cubic inches in a cubic foot) gives 77 6/9 cubic feet. Now that we have shown the practitioner how to determine the fall and the water quantity (required for designing a mill), in the following chapter, we will describe how to apply this knowledge.

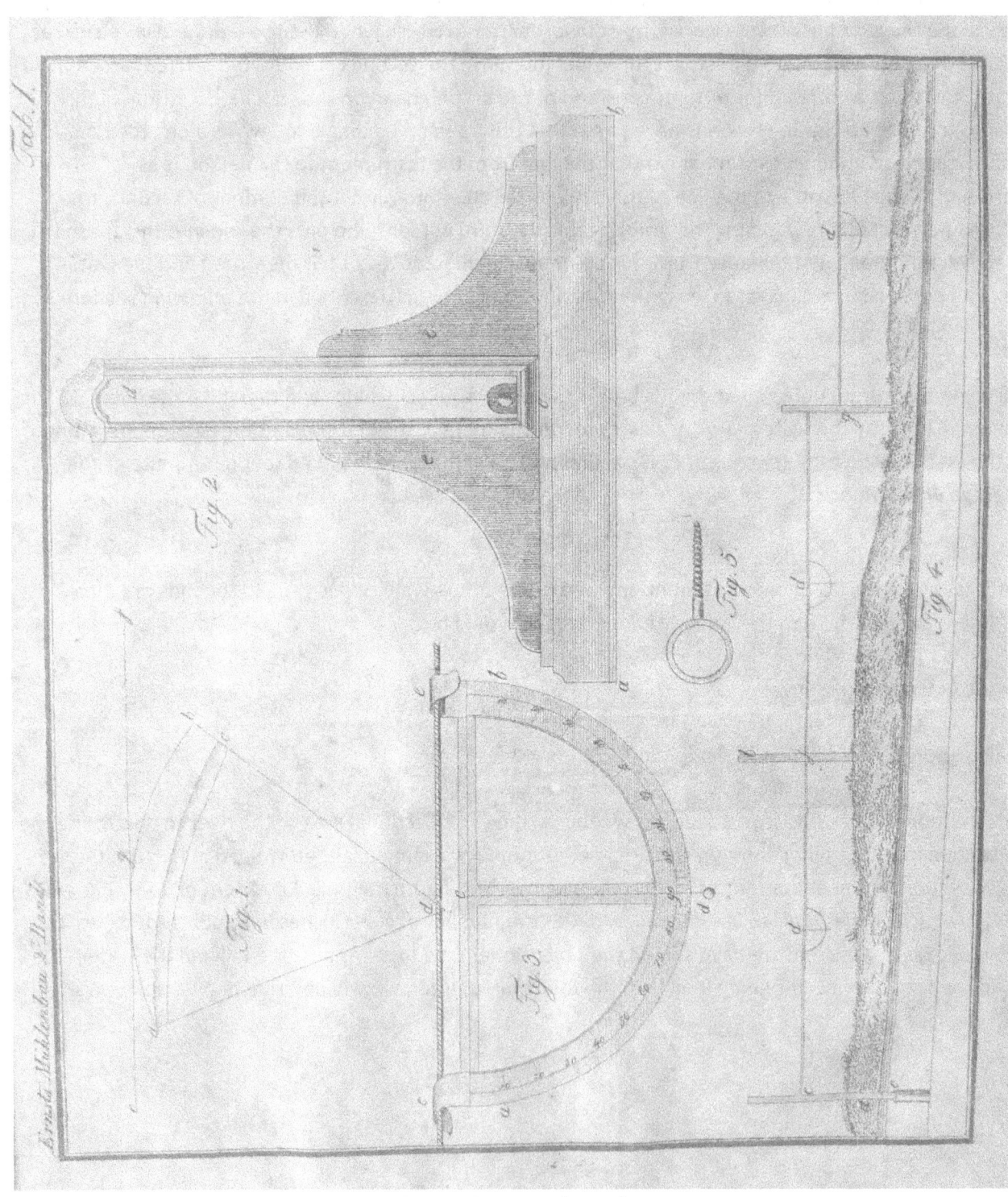

Plate I: Figures 1 through 5

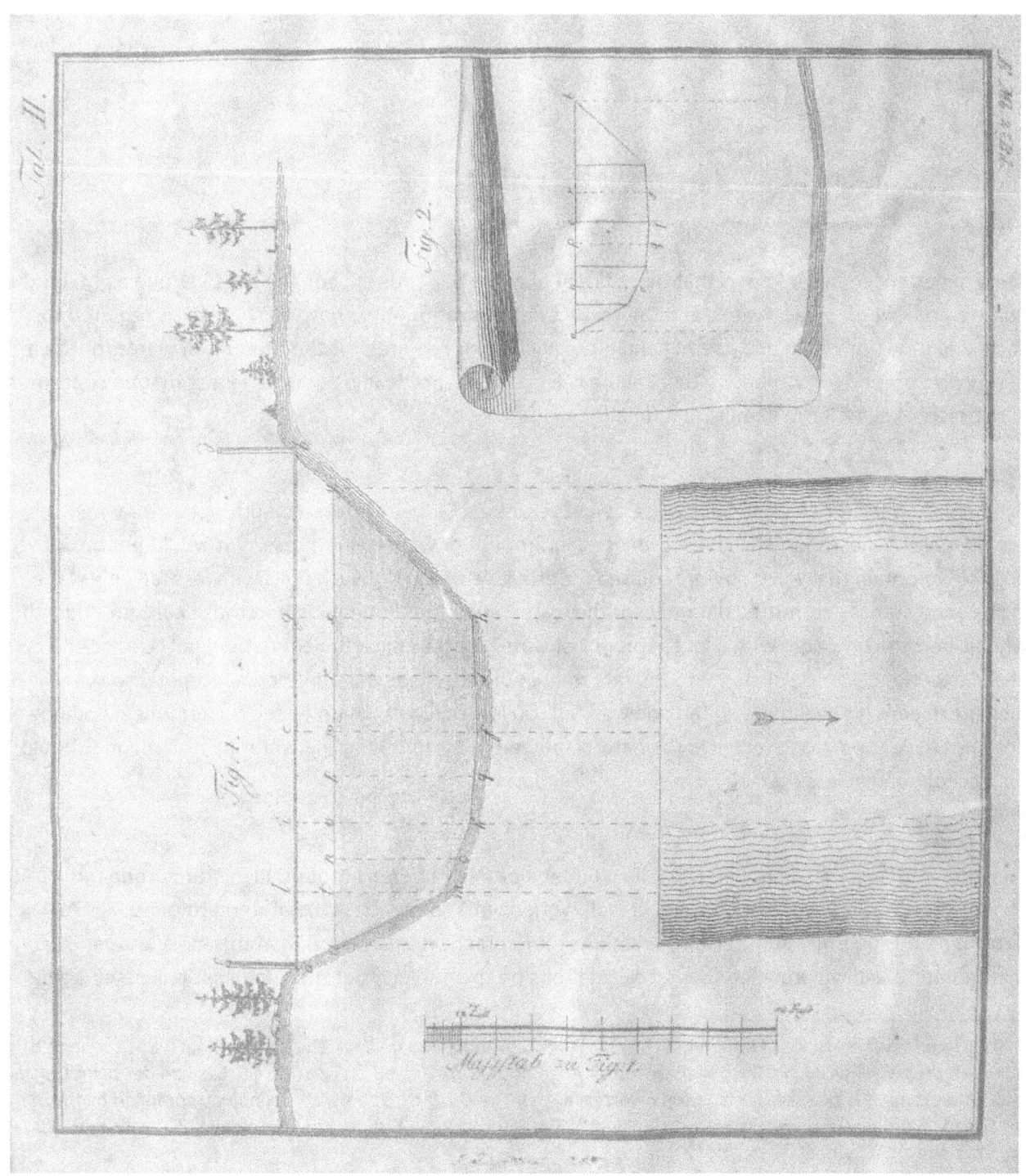

Plate II: Figures 1 and 2

Chapter 2

Building a Staber Mill[12]: Design and Practical Experience

Part 1

Every experienced miller knows that the efficacy of a mill depends only on the relative fall height, on the relative amount of impact water, and the correct gearing ratio. Alone how much water is required for each kind of water wheel and gearing and how much fall is required for this amount of water for the mill to function properly is something that can be calculated theoretically, but which must first be confirmed with practical experience in mills that function well.

Part 2

The minimum fall for an undershot wheel, whether it be a Staber or panster[13] (lifting) water wheel, should never be less than 10 inches. Even if a fair amount of water were present, it would not be possible to obtain the water power required for effective motion of such mills with less fall unless the local circumstances permitted damming of the water. Even then, one finds that these kinds of undershot wheels become unusable with a large amount of water, because insufficient natural fall is present. These experiences thus substantiate the fact that, in order for the mill to function properly so that the owner can earn his livelihood, when one wants to build a mill with an undershot water wheel and one does not have the right to increase the natural fall height by damming the water upstream, one should not accept a fall of less than 10 inches.

Part 3

Knowing from experience how much fall an undershot mill requires is not enough. Rather one must also observe how much water per unit time a well-working mill requires for the aforementioned fall. To determine this quantity with certainty requires more practical experience than theory. Various mathematicians have provided us with calculations on this subject, but these calculations never agree

---

[12] Translator's note: A Staber water wheel has two rims (shrouds) and no soal. The blades (floats) are connected between the two rims and lie flush with the outer edge of the rim. (based on Benzler, GS. Lexikon der beim Deich- und Wasserbau ... vorkommenden ... Kunstwörter und ...) The *Draft Dictionary of Molinology* compiled by the Dictionary Working Group of TIMS. Herts (England): The International Society of Molinology, 2004 calls it a "shroud-and-float" wheel.

[13] Technical annotation: Chapter 6 clarifies the distinction between a Staber mill and a panster mill. A Staber mill is a water-powered gristmill where one water wheel drives one pair of millstones. A panster mill requires about twice the water flow as a Staber mill, and its water wheel drives two pairs of millstones. In addition another mechanism is incorporated in a panster mill to permit the raising and lowering of the water wheel and axle to accommodate changing water levels in the adjoining stream.

with practical experience. For example, Mr. Karsten (a senior civil servant) puts the water quantity required by a Staber mill at 41 2/3 cubic feet per second for a 42-inch fall. If, however, a river supplies 41 2/3 cubic feet with the aforementioned fall, then one could build an average panstermill without hesitation and would not have to consider a Staber wheel with this amount of water and fall. Similarly, Professor Busch, in his hydraulics text, Section 40, assumes 16 ½ cubic feet of water per second for a Staber mill. On the other hand, Beyer assumes 11 2/3 cubic feet. These values may work in theory, but [in practice] a Staber mill requires 50 cubic feet of water per second with 10 inches of lively fall and 12 cubic feet per second with 42 inches of fall. If the measured velocity of a river is 50 cubic feet per second with 10 inches of lively fall, then a Staber mill can be built without hesitation. This mill could finely grind 10 to 12 Dresdner bushels of grain in 24 hours, which would allow any miller to earn his livelihood without cheating his customers.

Part 4

Now, based on practical experience, when one knows how much water per second (to achieve the aforementioned effect) a Staber mill requires, it is possible to use the rule of proportions to calculate the water quantity for different falls, or, conversely, the fall from the water quantity, as shown by the following examples. Let's say that you have to build a mill on a river, and this river supplies only 30 cubic feet of water per second, but, fortunately, more fall could be acquired. The following question arises: How much fall is required for 30 cubic feet of water, if a Staber mill is to function like the one described in Part 3? To answer this question, one infers that the ratio of 30 cubic feet to 50 cubic feet is the same as that of 10 inches of fall to the unknown fall.[14]

Formulation:
30 ft^3 : 50 ft^3 = : 10
        10
\-
       500

[Long division on p. 30 of German original]

16 2/3 ft of fall is required when there are 30 cubic feet of water and the aforementioned Staber mill is to grind grain as with 10 inches of fall and 50 cubic feet of water.

---

[14] Clarification: Technical annotation: An undershot water wheel gets its power from the rate at which kinetic energy is available from the flowing water as it impacts the blades. The following formula for the maximum power available to an undershot wheel in a flowing stream of water is derived from modern fluid mechanics. $P=62.4QH(1-\lambda^2)$ where P=power (ft-lb/sec); Q=flow(ft^3/sec); H=fall (ft), and λ= a dimensionless parameter related to the efficiency of the undershot wheel. Empirical studies of existing mills showed that the minimum practical value for $\lambda=1/2$. The values of Q and H shown in this chapter indicate that Ernst's design leads to a maximum available power of about 1950 ft-lb/sec or 3.5 HP. Since the available power is linearly proportional to both Q and H, the ratios that Ernst gives for equivalent situations hold. That is $Q_1H_1=Q_2H_2$ and therefore $Q_1:Q_2$ as $H_2:H_1$ for equivalent situations 1 and 2.

1) Note

It is not correct to assume that a mill's efficiency can be further increased by increasing the fall (when the conditions allow) beyond the value given by the aforementioned calculations when only a small amount of water is available. This calculation alone is necessary, when building a mill, to know what kind of water wheel and machinery can be used for different water quantities and falls and not, as often happens, just to leave it to chance whether the mill might work or not and the owner might make it or not. And it's all the same to many millwrights. On my trips, I've encountered many mills where the water quantities and falls called for an average Strauber[15] water wheel, but a 2 ells, 12 inch-wide mill race for a Staber wheel was built instead. How can such messed-up mills do what they are intended to do? For this reason, calculations based on experience and testing are necessary to guide the practitioner down the right road.

Part 5

In the previous section it was shown how the fall can be calculated from the measured water quantity. Conversely, one can calculate the incoming water quantity from the measured fall. Let's say that the river on which you want to build a mill has a lively fall of 2 feet or 24 inches over the prescribed leveled course. To determine the corresponding water quantity for this fall, one sets up the following ratio: 24 inches of fall to 10 inches of fall is the same as 50 cubic feet of water to the unknown water quantity.

Formulation:
$24 : 10 = : 50 \text{ ft}^3$
10

500

[Long division on p. 32 of German original]

20 5/6 cubic feet of water is required for 24 inches of fall.

If the river supplies this amount of water for this fall, then a Staber wheel can be used without any problems and the mill will function as intended.

Part 6

It is, however, not enough, when building a mill, to make calculations based on the measured water quantity and fall according to practical formulations. Rather, one has to know how to calculate the width

---

[15] Translator's note: A water wheel in which "mortises were cut into the rims of the wheel, starts were inserted in the mortises, and the blades were pinned to the starts" (Reynolds TS. *Stronger than a hundred men: A history of the vertical water wheel*. Baltimore: Johns Hopkins University Press, 1983, p. 163 and *Draft Dictionary of Molinology* compiled by the Dictionary Working Group of TIMS. Herts (England): The International Society of Molinology, 2004)

of the mill race from the determined water quantity and the established mill laws concerning the normal water depth (Standwasserhöhe??). In order to achieve this second goal, the practitioner must use practical experience if he is not to err from the correct path.

The average normal water depth is commonly assumed to be 3 feet or 36 inches for Staber and panster water wheels according to most mill laws in Saxony, as well as in other German states. However, there are also mills where the normal water depth is only 2 feet. Of course in these situations, the master millwright must choose the normal water depth according to his judgment when he builds a mill, when the laws do not permit it, according to the mill laws on site to avoid the unfortunate lawsuits involving mills, which throw entire families into poverty. Experience teaches that when a river supplies 50 cubic feet of water in 1 second and has 10 inches of fall, according to Part 3, with a normal water depth of 36 inches, a mill race for a Staber wheel has to have an inside width of 2 ells, 12 inches. This opening gives a cross-sectional area of 1800 square inches for the 50 cubic feet of water per second to pass through when the water depth is 36 inches.

After finding the normal water depth and the water quantity, one has to determine the width of the channel as follows: One assumes that the relationship between 50 cubic feet of water and a cross-sectional area of 1800 square inches is the same as that between the water quantity, for example, 40 cubic feet, to an unknown cross section.[16]

Formulation:
50 ft^3 : 1800 sq. in. = : 40 ft^3
40

72000

[Long division on p. 34 of German original]
The cross-sectional area of a channel for a Staber wheel has to be 1440 square inches when a river only supplies 40 cubic feet of water.*

---

[16] This section on race design reveals a major inconsistency with the previous section on water flow rates.

* Technical annotation: The water flow section was shown to be based upon an equivalence of available power between different water flow rate cases. The race design section can be shown to be based upon an equivalence of water velocity, and in fact it is a significantly lower velocity than that implied in the power section. The cross-sectional area recommended for the race is referenced to empirical data. Given that the flow rate is now known to be equal to the water velocity times the cross-sectional area: $Q=VA$, the base line flow rate of 50 ft^3/sec and recommended cross-sectional area of 1800 in^2 results in a base velocity of 4 ft/sec. The equivalence employed by Ernst to determine other cross-sectional areas is based upon keeping the velocity constant at 4 ft/sec. Hence, $V=Q_1/A_1=Q_2/A_2$, and $A_2=Q_2A_1/Q_1$. This results in the recommended $A_2=1440$ in^2. Note: The maximum available power written in terms of water velocity from modern fluid mechanics is $P=0.97\ QV^2(1-\lambda^2)$ where P(ft-lb/sec); Q(ft^3/sec); V(ft/sec), and λ is again ½. For a flow rate of 50 ft^3/sec and a velocity of 4 ft/sec, the resulting available power is a maximum of about 582 ft-lb/sec or 1 H.P. This is considerably below the previous available power of 3.5 H.P. The reason is that the desired 10 inch fall of the previous section leads to a theoretical water velocity of 7.3 ft/sec rather than the design race velocity of 4 ft/sec. This difference could be ameliorated if the 10 inch fall was intended to occur in a chute right before the water wheel at the end of the head race.

When one now divides the normal water depth into the cross-sectional area, the quotient gives the width of the channel. For example, dividing 1440 by a water depth of 36 inches results in 40 inches, or 1 ell, 16 inches. This is how wide the channel has to be for 40 cubic feet of water and a depth of 36 inches.

However, should the mill laws only allow a normal water depth of 30 inches, and the river supplies 50 cubic feet of water in 1 second, how wide would the channel then have to be? One proceeds as described above, in other words, one divides the cross-sectional area corresponding to 50 cubic feet (= 1800 sq. in.) by the depth of 30 inches. The resulting quotient of 60 inches or 2 ells, 12 inches is the maximum width of the channel for the Staber wheel. This procedure would be applied to each normal water depth.

Part 7

Now that we have explained the initial prerequisites concerning the water, which the practitioner must observe, we can turn our attention to constructing the mill itself. We will show the rules and regulations concerning mill construction. Before a master millwright decides to build a mill and even when he is legally obligated to build one, he must still familiarize himself with the natural possibilities, which are based on the available, suitable fall and the relative head of water in addition to a good location.

Part 8

A good location is one that is not only not too far from the surrounding or authorized customers, but also one that has a lively head of the required water. Sometimes there are situations in which the master millwright is ordered to build a mill where a mill has been standing for ages. The millwright should regard this order as a warning, because there must have been some reason why the previous mill failed. He must try to discover the reasons, to determine if somehow these could also have a negative effect on his building plans or if these problems could be overcome by judicious planning. There are many different reasons why a mill could have failed. Perhaps the mill that failed was having a negative effect on the nearby mills with regard to the fall. In other words, perhaps several mills were located so close together because the millwright didn't know any better, resulting in each mill not functioning as it should due to insufficient fall. By taking away the fall used by this one mill, the neighboring mills could be converted to functioning mills. Perhaps some mills had been built at a site where the adjacent properties were flooded as a result of damming the river and causing excessive fall. Those kinds of mills do more harm than good. Or was the problem due to faulty equipment, causing the customers to leave? All of these possibilities and more must be carefully investigated by a sensible master millwright before he begins constructing a new mill. Because what good is a good mill when the economic plan has not been correctly thought through? An adequate fall and a reasonable head are the second important considerations when building a mill. All rivers and streams have fall, but not necessarily enough to allow a mill to function well. Undershot water wheels don't need as much fall, but require more water. Overshot wheels, on the other hand, require less water, but more fall. Strauber and flutter wheels are somewhere in between. Thus after all of the aforementioned circumstances have been carefully investigated, one must know what kind of water wheel is appropriate for each fall and water quantity.

Part 9

When planning the construction of the mill itself, after all of the aforementioned circumstances have been investigated carefully, leveling the fall is the main thing that must be done with the highest precision. Where a mill is built is either determined by a certain course of the river from which the fall is taken away or, when circumstances allow, to choose this course yourself, as described next. At the site where the mill is to be built, one inspects the height of the bank from this location upstream to the end of the prescribed or selected course of the river to find the lowest point. In this process, one should not worry about small plots of land that can be protected easily from flooding, as a result of the higher water level, with a dam. Rather, one must be concerned with entire fields and meadows, which cannot easily be protected from flooding. From this lowest lying point to the selected site of the mill, level not the fall of the water, but the fall of the bank. After doing so, if you find that the height from the lowest point of the bank to the surface of the water is not 2 to 3 feet, then you should not go to the trouble of building an undershot mill if you are worried about flooding the properties or even cities and villages, which, unfortunately, has happened at many locations. The ultimate goal of leveling a bank is none other than to determine if the relative normal water depth of the river can be increased by one or two feet at the site where the mill is to be built without flooding the area around the lowest lying bank. If the lowest point of the bank is higher than 2 or 3 feet, then the mill can be built even more advantageously. However, we're not talking about exceptional floods here.

Part 10

The second order of business when building a mill concerns determining the head of water, which, as indicated in the previous chapter, has to do with measuring the "consumption." This procedure was explained in Chap. 1, Part 11 u.f.w.??.[17] However, the following must be mentioned when building a mill: When measuring the water quantity, don't choose the greatest, instead choose the usual average water depth of a river. If you choose the usual smallest water depth, then most of the mills could not be built to achieve the required purpose. If you choose the greatest depth, then the mill would only function during times of high water. If you determine beforehand how much water a river supplies in one second, then you also know, based on the practical formulation given in Part 3, if the river provides enough water to build a Staber mill.

Part 11

Let's assume that, after leveling and measuring the water quantity, you find that a river has 12 inches of fall and delivers 45 to 50 cubic feet of water per second over a certain course. Under these conditions, a Staber mill could be built, which promises to give a good result. The first rule concerning a Staber mill, given a suitable fall and water quantity, is to determine the height of the water wheel and the corresponding gear ratio of the millstone. The height of the water wheel should be determined, in accordance with the rules of mathematics and physics, by the force of the water on the impact surface

---

[17] Translator's note: I have not been able to find the exact translation of this abbreviation, but from context it seems to mean something like "and subsequent pages."

of the blades. Alone because this calculation for determining the height of the water wheel requires scientific knowledge, most of my readers would not be able to apply it. Furthermore, this calculation is seldom useful in practice. A Staber wheel, which puts a mill in motion, is seldom higher than 9 ells and or lower than 6 ells in practice. A low fall requires the water wheel to be higher than a large fall, but there are also limits when it comes to this kind of wheel. Experience shows that the fall for this kind of wheel should never be less than 10 inches and, in the second case, may not be more than 30 inches if one expects the mill to function well with an open Staber wheel. A fall of 10 inches requires a water wheel that is 8 ells high when the water volume is 50 cubic feet. If the head were not 50 cubic feet per second, but 45 instead with the same 10 inch fall, then it would be better to make the wheel 9 ells high. For Staber and panster mills, in practice, as the fall increases from 6 to 6[18] one decreases the height of the water wheel by one ell. For example, if the fall is 10 inches and the water quantity is 50 cubic feet, the water wheel should be 8 ells high. If the fall falls between 10 and 16 inches, then the water wheel should be 7 ells high. If the fall is between 16 and 24 inches, then the water wheel is 6 ells high. In any case, the fall must deliver the required water quantity, which can be determined easily using the calculation described in Part 5. If the fall is more than 30 inches, then the Staber wheel does not work well. Rather, when the river supplies only about 12 to 14 cubic feet of water per second, a Strauber wheel is used. If, however, the river supplies 18 to 20 cubic feet, then a flutter wheel is best. In this manner, the height of Staber and panster water wheels can be determined by the practitioner, which also works out best in practice.

Part 12

After determining the height of the water wheel based on the measured fall and the required water quantity, the next thing we have to do is to determine the rotation of the millstone relative to the particular height of the water wheel. Again, we have to let experience guide us. Observations of well-functioning Staber mills show that a Staber wheel that is 8 ells high, with 10 inches of lively fall and 50 cubic feet of water, when the mill is running at full capacity, completes its rotation in 6 seconds when the millstone, which has a diameter of 1 ell, 12 to 14 inches, has completed 12 rotations. Thus, it completes 2 rotations per second. This practical formulation concerning Staber wheels can be applied to different water wheel heights to easily determine the rotations of a millstone using the rule of proportions, as shown clearly by the following example. Let's say that the fall and the associated water quantity require the water wheel to be 7 ells high. The question we can ask is: How many rotations would the millstone (with the aforementioned diameter) complete with that water wheel height? One proceeds as follows: The ratio of an 8 ell-high to a 7 ell-high water wheel is the same as the ratio of 6 seconds to the unknown time of rotation that is produced by a 7 ell-high wheel.

Formulation:

8 : 7 = 6

6

42

---

[18] Clarification: Technical annotation: In other words, from the baseline fall of 10 inches, each successive span of 6 inches increase of fall allows for an additional decrease of 1 ell in wheel diameter.

[Calculation on p. 43 of German original] 5¼ seconds[19]

Doubling this time, namely 5¼ seconds, gives the rotation of the millstone, which is 10½ rotations with a 7 ell-high water wheel. This is the procedure that can be used for every height of a Staber water wheel. The calculations for the rotation of millstones in Staber mills that have several water wheels in one channel will be described elsewhere. The aforementioned calculation applies only to one water wheel that has its own channel.

Part 13

The gear ratio of the pit wheel[20] to the water wheel varies greatly, but experience with well-functioning Staber mills has shown that the best ratio of pit wheel to water wheel is almost always 1 to 2. Some mills use a ratio of 2 to 3 or 3 to 4 for the size of the pit wheel to the water wheel, but these ratios are usually used in situations where the water quantity or fall required for a Staber wheel is not quite sufficient. With these ratios, the water wheel turns more slowly, so that the water has time to accumulate ahead of the blades and thereby provide the mill with a little more power. However, the 1 to 2 ratio of pit wheel to water wheel remains the best, not just for Staber wheels, but also for other simple undershot water wheels (assuming, of course, there is sufficient water quantity and fall for each kind of wheel, based on the aforementioned rule). Careful observers will have noted this in practice. Regardless of the type of water wheel, if you want to make the water wheel turn more slowly when less water is available, but still maintain the rotational velocity of the millstone, change the gearing. This can easily be done, when there is a little or a lot of water, by pushing just one gear forward[21], which has one less cog than that suited for the normal water level. That is why it is wrong for Beier, Melzer and others to have treated this important rule so superficially in their mill books, only always for Staber and other wheels. For example, for Staber wheels at most water levels, they assume only 6-stave gear wheels and wide spacing.

The best gear ratio for a Staber mill with 10 inches of fall and 50 cubic feet of water per second is as follows: Select an 8 ell-high water wheel (as mentioned above) and give the millstone 12 rotations. This

---

[19] Technical annotation: A study of Ernst's example reveals that the underlying equivalence in determining the rotational speeds of water wheels of different diameters is maintenance of the tangential rim velocities, say V_t. From modern fluid mechanics, $V_t = R\omega$, where R is the wheel's radius, which is ½ its diameter, and ω is its rotational velocity given in radians per second. Since one revolution equals 2π radians, ω can be measured in revolutions per second for the purpose of applying Ernst's ratios. Thus, for two equivalent cases: $V_{t1} = R_1\omega_1 = D_1\omega_1/2 = V_{t2} = R_2\omega_2 = D_2\omega_2/2$. And $D_1\omega_1 = D_2\omega_2$; $\omega_2 = D_1\omega_1$ and the time for one rotation of the second water wheel is the reciprocal of ω_2. For the case of $D_2 = 7$ ells, the result is 5 ¼ seconds. Note that since the smaller diameter wheel was previously determined to be suitable for cases of larger falls, the assumption of equal tangential rim velocities could be $D_8/D_7 = 1/\omega_8 : 1/\omega_7 = \omega_7/\omega_8$.

[20] Translator's note: A vertical, cogged wheel. Called "trundle wheel" by Holt R. The Mills of Medieval England. Oxford (UK): Basil Blackwell Ltd. 1988.

[21] Translator's note; Original German text – "nur ein Getriebe vorstößt".

is achieved with a pit wheel having 83 cogs (teeth, wooden pegs) spaced 3 ½ inches apart, which meshes with a 7-stave gear wheel, but with the proviso that the millstone retains a diameter of 1 ell, 12 to 14 inches with this ratio. Based on this ratio, the number of cogs can easily be determined for different water wheel heights, not just from the ratio of the number of rotations of the millstone, but also using the following calculation. For example, as shown in the previous section, if a millstone completes 10½ rotations with a 7 ell-high Staber water wheel, how many cogs would the pit wheel have to have with a 7-stave gear wheel to get this many rotations? The formulation is as follows: The ratio of 12 rotations to 83 cogs is the same as 10½ rotations to the unknown number of cogs.

Formulation:
12 : 83 = 10½
----
24 21   21 halves

       83
     166

   1743

[Long division on p. 47 of German original]
72, which means that one assumes 73 cogs for this pit wheel.

Another method for determining the number of cogs for the pit wheel is based on the measured rotation of the millstone: Multiply the number of staves by the number of rotations of the millstone. The resulting product gives the required number of cogs for the pit wheel.

Part 14

One important aspect of constructing a mill, which deserves our attention, has to do with different sizes of millstones, which has to do with the established milling process. A state or a region in which most of the grain is roughly ground and sifted and not processed, longer stones can be used than in regions where most of the grain is finely sifted and processed. For example, in our region around Dresden and Leipzig, most of the grain is finely sifted and processed by means of very fine bolting bags. That is why our stones cannot be larger than 1 ell, 12 to 16 inches in diameter and 20 to 26 inches high if we want to prevent pasting. In most Prussian states and in Sweden and other northern countries, where most of the grain is only dry-ground, larger millstones are used. The diameter of those millstones is 1 ell, 18 to 20 inches for Staber water wheels and over 2 ells, 4 to 6 inches for panster wheels, but they are only 18 to 20 inches high. A millstone with a 1 ell, 12 to 14 inch diameter and which is 22 inches to 1 ell high weighs 14 to 15 zentner,[22] depending on the mass of the stone. On the other hand, a stone with a 2 ell, 4 to 6 inch diameter and which is 18 to 20 inches high weighs 20 to 22 zentner. The latter is just about the ratio of size to weight of the millstones in our German watermills. Let's say that you want to build a

---

[22] Translator's note: One zentner = 100 German pounds. One zentner is about 50 kg (Wikipedia).

Staber mill in those regions where grain is only roughly ground, where the millstone could have a diameter of 1 ell, 20 inches. We could calculate the number of rotations of that kind of millstone using our rule of proportions as follows: The ratio of the diameter of the long stone = 44 inches to the diameter of the short stone = 36 inches is the same as the proportional rotation of the short stone = 12 to the unknown number of rotations of the long stone. The unknown = 9 9/11.

Formulation:
44 : 36   = 12
     12

   72
   36

  432

A millstone with a diameter of 1 ell, 20 inches would have 9 9/11 rotations with an 8-ell high Staber wheel.[23]

   1) Note

The rotation of millstones with diameters larger than those allowed by the ratios based on experience can also be calculated using this method. One may not have a millstone that is 2 inches bigger than the ratio allows without adjusting the rotations, otherwise the mill comes to a halt and the stones become uneven and collide, with negative consequences. One also spaces the cogs on the pit wheel ½ or ¾ inches farther apart, for example, 4¼ inches, and uses fewer cogs while adding a couple of staves more to the gear wheel. By doing so, the gear ratio is correct and the ratio of the pit wheel to the water wheel is nearly 1 to 2. If this rule is observed, one can be sure to expect good work from a Staber mill.

Part 15

In the previous section, I described the rules for determining the height of the water wheel and the gear ratio for the millstone. Now let's turn our attention to the layout and dimensions of the individual components of a Staber mill with one millstone. I will describe the rules for building this kind of mill and their practical advantages. The first task is to determine the length of the weir sill I[24] (Plates III, IV, and V). To do so, make a diagram and mark the sill according to the following dimensions.

---

[23] Technical annotation: Ernst's proportional relationship for calculating the rotational speed of a given millstone is based upon the equivalence of the tangential speeds at the rims of the differently sized millstones: Therefore $D_1\omega_1/2 = D_2\omega_2/2$.

  Note: In this section the diameter of the long stone is given as 1 ell, 20 inches or 44 inches. This indicates that 1 ell = 24 inches as compared to 19 inches as determined from Chapter 1, Section 6. Thus a discrepancy exists. The dimensions given for the following mill structures are based on the 24-inch ell.

[24] Translator's note: In Part 17, the sill is labeled "G" and the apron is labeled "I". On Plate III, "G" seems to be the sill and "I" the apron.

	Ells	Inches
Depending on conditions, the sill goes from the sluice post H, Plate III, into the bank	3	---
For the width of the sluice post H	---	18
For the width of the spillway, here	2	---
For the width of the second sluice post	---	18
For the width of the wheel race	2	12
For the width of the third sluice post	---	18
From the third sluice post to the wall M	1	
For the position of the sill behind the wall	2	---
Sum for total length of the sill	12	18

The second group of adjacent components visible on the horizontal plane has to do with the water wheel shaft as follows:

	Ells	Inches
For the neck of the shaft at E	---	20
For the width of the head stock F	---	12
For the width of the side wall L	---	6
Sum for transport	1	14

	Ells	Inches
For the space between the shrouds and the side wall	---	1
For the width of the water wheel A including the shrouds	2	---
For the second clearance of the shroud	---	1
For the width of the second side wall	---	6
For the width of the second head stock	---	12
For the space between the head stock and the wall M	1	3
For the thickness of the wall M	1	---
For the space between the wall and the sill N	---	6
For the width of the sill N	---	18
For the space between the sill and the pit wheel B	---	15
For the width of the pit wheel B	---	14
For the space between the cross piece Q from the pit wheel to the neck of the shaft	---	18
For the length of the neck	---	20
Sum for total shaft length	10	8

All adjacent components having to do with the water wheel shaft are laid out according to these dimensions.

The third group of components relates to the length of the building as follows:

	Ells	Inches
For the width of the hursting[25]	4	---
For the length of the bolting hutch T	3	---
For the space between the bolting hutch and the wall	3	6
For the wall thickness	1	---
Sum for length of the building including both walls	**11**	**6**

The fourth group of components relates to the width of the building as follows:

	Ells	Inches
For the wall thickness M	1	---
For the space between the wall and the pit wheel B	1	12
For the height of the entire pit wheel, which has 73 cogs spaced 3½ inches apart	4	1
For the space between the pit wheel and the wall	2	12
For the wall thickness	1	
Sum for width of the building including the walls	**10**	**1**

2) Note

When calculating the width of the building, one should actually also note the distance between the bridge tree O, O and the brayer beam P, P[26]. This value is easily determined from the center line of the pit wheel, so it's not necessary to mention it when there is only one pair of millstones. However, with multiple pairs of stones, these dimensions must be included on the drawing due to the position of the shafts.

Part 16

After determining the position of the adjacent parts based on their dimensions in the floor plan, we turn our attention to their relative position in the vertical plane, from which the height of the hursting and the building is determined. If we go in order, this would be the fourth[27] group of components that has to be measured. Starting at the cross-tree a in Fig. 1 on Plate V, calculate the dimensions using Plate IV as follows:

	Ells	Inches
For the radius of the water wheel	3	12
For the radius of the pit wheel	2	½
For the space between the pit wheel B and the post of the hursting	---	14
For the width of the post of the hursting	---	4
For the height of the millstones	1	12
For the height of the hopper	1	18
For the space between the hopper and the roof beams	1	3
Sum from the cross-tree (as the lowest point of fall) to just below the roof beams	**10**	**15½**

The sills N N are laid 16 to 18 inches lower than the center of the water wheel shaft and this position also determines the position of the floor. These are the main components and their dimensions with

---

[25] Translator's note: The *Draft Dictionary of Molinology* uses the term "hursting" for Mühlgerüst, literally "mill framework."
[26] Translator's note: In Part 17 and on Plates III and IV, the brayer beam (Tragebank) is designated P R.
[27] Translator's note: I think he means "fifth."

which you must become familiar when accepting a contract to construct a Staber mill, so that you can give the workers the correct instructions. With this information, the work will go faster than if you have to spend half a day determining the size and position of every component.

Part 17

I have just described the components that a master millwright must know about when constructing a Staber mill with one pair of millstones. Now it is my duty to describe the other parts – their names and how they are laid out. The floor plan in Plate III shows the following parts to scale:

A	The water wheel, which is 7 ells high and has 32 blades on its periphery
B	The pit wheel, which has 73 cogs spaced 3½ inches apart, which meshes with a gear wheel with 7 staves. With this gearing, the millstone revolves 10 3/7 times for each turn of the water wheel.
C	The water wheel shaft
D D	The beams beneath the pillow block, with the pillow block E
F F	The head stocks
G	The sill
H H H	The sluice posts
I	The piles of the apron
K	The posts in the mill race
L L	The side walls
M M	The foundation of the masonry
N N	The sills of the hursting
O O	The bridge tree
P R	The brayer beams
Q	The cross piece upon which the bearing rests, in which the spindle rotates
S S	The solepieces on which the sills partially rest and are attached to
T	The bottom of the bolting hutch
U	The pillow block for the pit wheel

V	The steps up to the hursting
W	The mill room[28]
Z	The door

These are the components that can be seen in the floor plan. They are represented by the same letter in the top view as in the side view, where the components are shown in cross section. Plate IV shows the side view from the Z F line in the floor plan. The following components, which were not visible in the horizontal plane of the floor plan, are shown in the vertical plane in the side view:

a	The hopper
b[29]	The front posts on which the hopper rests
c	The rear stilt
d	The millstone
e	The tun[30]
f	The longitudinal hurst beam
g	The skirting board
h	The posts of the hursting
i i	The braces that are mortised into the longitudinal hurst beam f and the bridge tree O O.
k	The breastsummer
l	The shaft together with the stop
m	The jog-scry in which the shaft is positioned
n	The gear wheel[31]
o	The arm from which the hopper is suspended.

Plate V, Fig. 1 shows the profile from the water side and Fig. 2 the profile through the hursting. I want to point out the following components, which are visible in the vertical plane: In Fig. 1, a is the cross-tree

---

[28] Translator's note: a small waiting room for customers or a room where journeymen could sleep
[29] Translator's note: This part was mislabeled "d" in original German text.
[30] Translator's note: Also called vat, wooden case, stone cover, drum, and hoop in *Draft Dictionary of Molinology*
[31] Translator's note: Holt (1988) calls this gear wheel a lantern pinion. Technical annotation: In other references it is also called the wallower.

that lies below the center of the shaft of the water wheel and which denotes the lowest point of the fall; in this case it is 12 inches from the top of the sill I. b is the sub-sill into which the sill is lap-joined. It serves to stabilize the sill against pressure from the side. Two to three similar sub-sills can be laid perpendicular to the sluice; this topic will be described elsewhere in more detail. The layout and scale of the rest of the sills and posts can be seen clearly in the diagram.

Figure 2 shows the section through the hursting with the bolting hutch removed. This allows the following components to be seen: b the beater shank along with the beater head Q and the shaker r, which produce the motion of the bolter. P is the longitudinal hurst beam head, which supports the lifting arm that positions the stones. The lifting bar passes through the stones and is attached to the head of the brayer R below. Even beginners, who only work as laborers in mill construction, know the purpose of each of these parts and why they are required in a mill. That is why it would be superfluous to say anything about the many details.

Part 18

I cannot fail to say something about a few practical rules that the practitioner should follow, which concern not just the correct determination of the dimensions but also the associated durability of various parts when constructing a mill.

1) The first dimensions when laying out a hursting are determined from the center line like this: We know from Section 17 that the cross-tree always lies below the center of the shaft of the water wheel. From the midline of the cross-tree and perpendicular to the front of the mill race we draw a center line, which forms the basis for laying out the hursting. Once we've determined the gear ratio, so that we know the height of the pit wheel, then we add another 6 inches over the half height of the pit wheel, starting from the middle, to each side. This cutout (Abstich??) shows the middle of the bridge tree.
2) If the hursting is to be durable and sturdy, so that it does not sway back and forth in the first year or often even earlier, as, unfortunately, experience has shown, the mortise and tenons not only have to be worked precisely, but the braces i i in Plate IV should not be lap-joined using the old method. Rather dovetail joints should be used to join the bridge tree and the longitudinal hurst beam.
3) The spindle, or the vertical line through the center of gravity of the millstone, must fall directly on the center of the pit wheel shaft to ensure that the stones rotate smoothly.
4) The water wheel shaft (onto which the Staber water wheel and the pit wheel are attached) must be positioned precisely so that it is perpendicular to the straight front edge of the mill race.
5) The bedstones must lie exactly in a horizontal plane if the mill is to work well. Dressing, or cutting the furrows, must be done correctly depending on each type of stone, which will be described elsewhere.
6) The hopper must be positioned so that the grain and groats flow uniformly under the stones through a fixed and even position of the shoe. To achieve this goal, I recommend using the roller-hopper invented by Mr. D. Melzer. With this device, one need not be concerned about

any sticking or clogging with any kind of groats, as often happens with shakers. I will reproduce and describe this kind of hopper at the end of this book.

7) The gear ratio for the shaker (or the bolter) must be proportioned correctly, otherwise the flour will never be separated cleanly from the groats and, when it hits the stones again, the flour flies everywhere and undergoes a kind of fermentation, which causes it to stick to the walls of the tun and bolting hutch. This results in the loss of the tasty and spiritual (geistig??) parts of the flour. The surest test for determining that the flour has been separated from the pulverized groats is to put some of the groats that have been sifted through the bolting bag in your hand and to blow down on it forcefully. If it flies cleanly off your hand and doesn't make or leave dust behind, then you can be sure that the flour has been separated cleanly. Three important pieces have to be in correct proportion to each other for the shaker to work well. These three pieces are: the position of the jog-scry, the position of the shaker arms, and the position of the beater head. To determine the correct position of the jog-scry, one proceeds as follows: Take the radius of the gear wheel. Forttragen?? this radius, starting from the middle of the pit wheel gudgeon 4 times on the sill. The intercept is the middle of the jog-scry.

If one now assumes 1½ radii of the gear wheel, determined from the center of the jog-scry (based on the position of the jog-scry), then the intercept gives the location of the gudgeon of the shaft. The beater arms must always be positioned horizontally. In this position, the beater arms must cut off (abschneiden??) the third part of the length of the bolting bag. If no shaking rod (Stemmrute??) is used with the shaker, where often the thumbs (Daumen??) and damsel rod burn off, then the beater head must be placed in such a position that it accelerates the elastic tension of the bag with its own weight. Because the usual length of a beater head is 1 ell, 4 to 6 inches, it must form an angle of 30 to 35 degrees with the horizontal beater arms. If the shaker is laid out according to these rules, the mill will fulfill its purpose even with regard to this important point.[32]

---

[32] Technical annotation: Item 7) here attempts to describe the mechanism used to drive the bolter (sifter). To better understand the directions as given, one would require a more detailed description of the bolter than that provided.

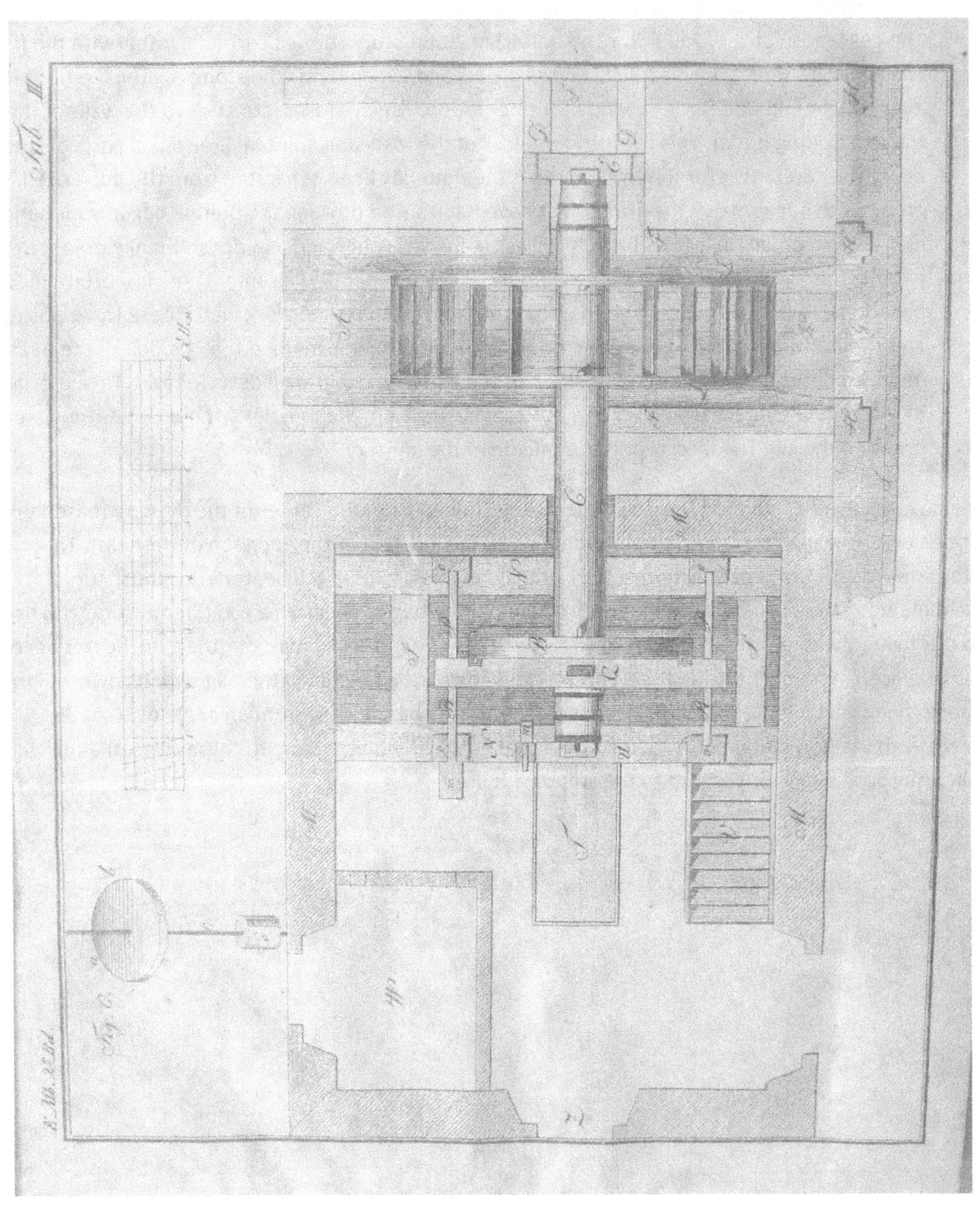

Plate III (Annotations on next page)

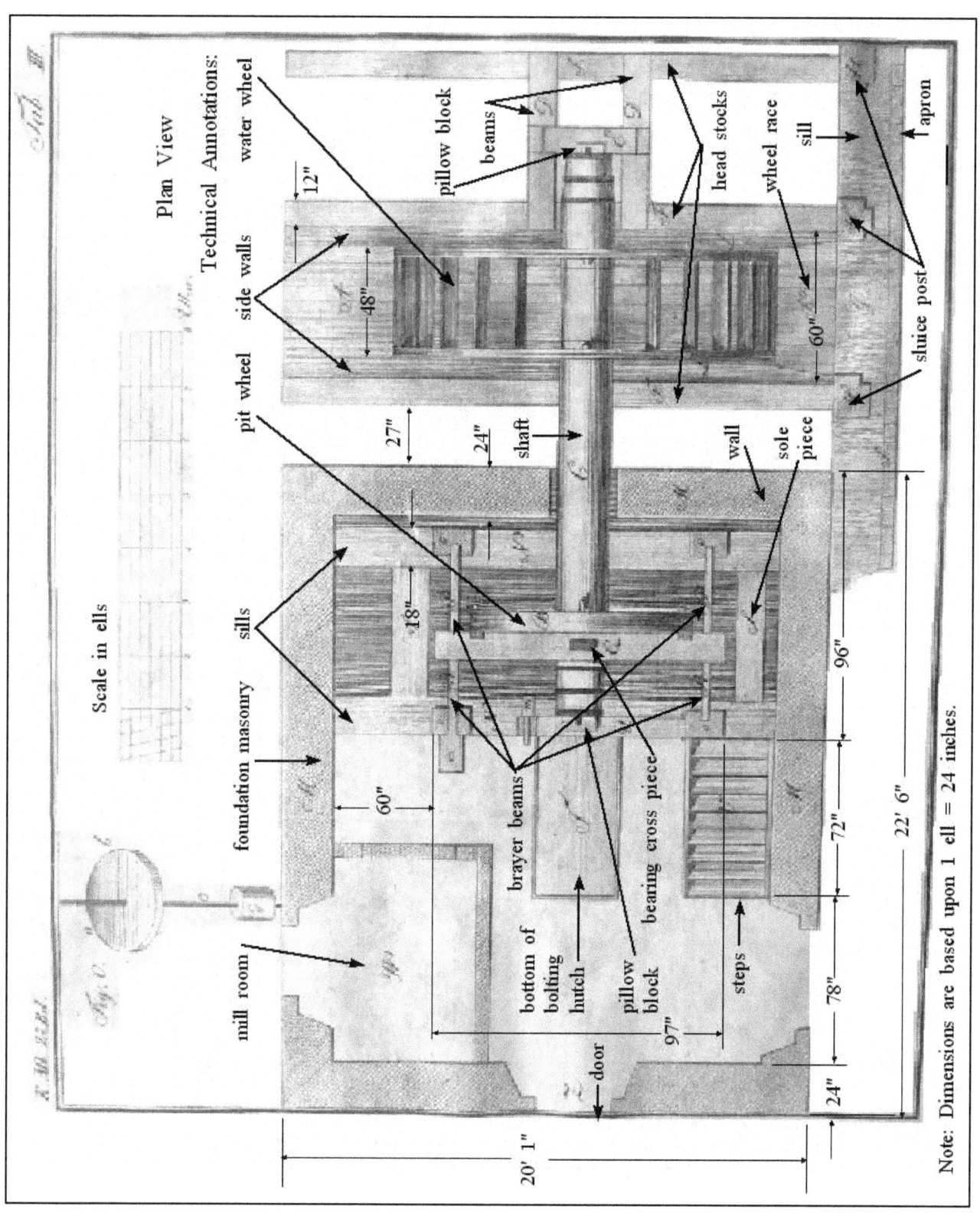

Plate III (with technical annotations)

Plate IV (technical annotations on next page)

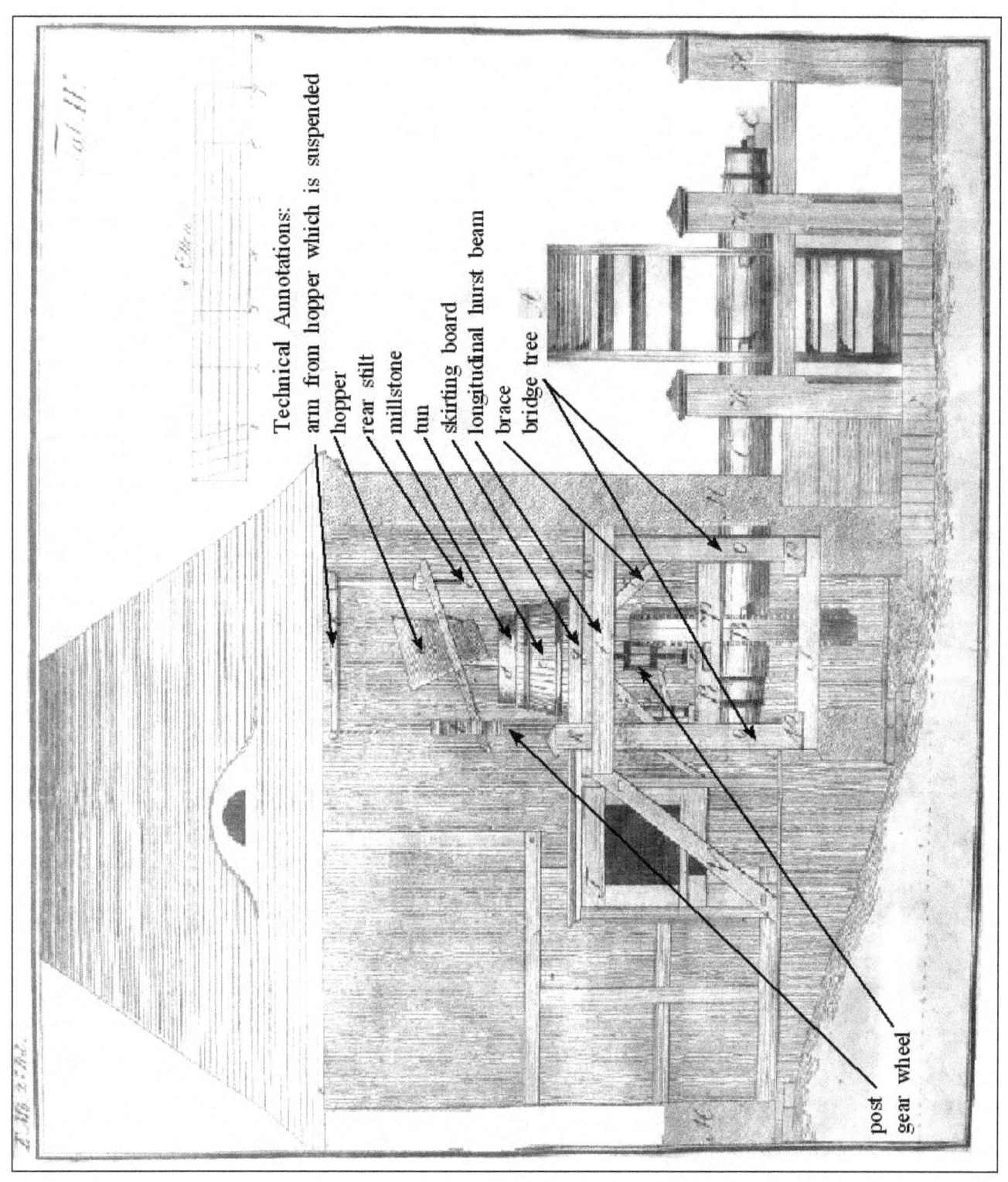

Plate IV (with Technical annotations)

Plate V

Chapter 3

Building a Staber Mill with 4 Millstones

Part 1

If a river possesses more water and fall than is required to build a Staber mill with one pair of millstones, then more millstones can be added. The measured fall and water quantity must, however, still be assessed precisely based on the aforementioned rules before construction begins. Let's say that a river possesses enough water and required fall to build two mill races and that two Staber water wheels could be hung in each channel. This kind of mill would be laid out according to the following rules:

1. Assuming that you found enough water and fall to build a Staber mill with two channels, then the main rule regarding such a mill has to do with how to divide the fall when two or three wheels are to be positioned in one channel.
2. Then all components are to be correctly positioned, both horizontally and vertically based on the specific height of the wheels, so that no component opposes the intended purpose.
3. One must also make sure that water does not hinder the construction of the sluices. In the following chapter, I will describe how sluices are constructed and use drawings to try to explain the design clearly to practitioners.

Part 2

The drawings shown on Plates VI, VII, and VIII show the design of a Staber mill with 4 millstones, designed for large gearing (großes Zeug??), for which the fall is 27 inches, and assuming that the required water quantity is provided. In order to divide the fall appropriately, so that the force of the water is distributed equally over each water wheel, proceed as follows: Partition the fall into 9 equal parts based on the uneven numbers 1, 3, and 5. The upper water wheels receive 5 parts, which corresponds to 15 inches of the total fall. However, one must remember to give the second water wheel, designated A in Plate VII[33], 2 inches of clearance (Risch??) from the weir sill b to the apron sill a and the remaining 13 inches for the fall from the apron sill a to the cross-tree c. The upper water wheel, on the other hand, is allotted the total 15 inches of fall, in other words, from the weir sill to the cross-tree. Furthermore, if one gives 2 inches of clearance (Risch??) from the cross-tree c in Plate VII of the second water wheel to the apron sill d of the fourth water wheel, then 10 inches of fall remain for the lower water wheels. One can still build without risk the cross-trees e of the lower water wheels 4 inches below the dead water, in which case the lower water wheels receive 14 inches of fall. The back pressure of the dead water standing on top of the lower cross-trees would be cancelled out without difficulty by the accelerated pressure of the standing water in front of the sluice gate. This is something that every miller knows from experience.

Part 3

In the previous chapter I described the position and dimensions of the gear wheels and the other components required to build a Staber mill with one millstone. When you add millstones, however, you have to dimension the components differently. Specifically, with several water wheels in one channel,

---

[33] Translator's note: The water wheels are labeled in Plate VI, not Plate VII.

different gear ratios must be considered. For this reason, it is my duty to show my readers in each diagram how the entire mill is laid out.

Part 4

To determine the length of the weir sill *a* on Plate VI, dimension the components for this Staber mill as follows:

	Ells	Inches
Depending on conditions, the sill goes into the bank	3	---
For the width of the first sluice post, which stands next to the bank	---	16
For the width of the spillway, depending on the river, here	4	---
Sum	7	16

	Ells	Inches
Sum carried forward	7	16
For the width of sluice post b	---	16
For the width of the channel for the first Staber water wheel, here	2	12
For the width of sluice post c	---	16
For the width of the second channel	2	12
For the width of sluice post d	---	16
For the space between the sluice post and the wall e	---	20
For the position of the sill behind the wall	1	12
Sum for total length of the sill	17	---

To calculate the length of the water wheel shafts, 2 short ones and 2 long ones, the second group of components is dimensioned as follows:

	Ells	Inches
For the length of the neck of the shaft c	---	18
For the thickness of the head stock f	---	12
For the space between the water wheel shroud and the head stock	---	6
For the width of water wheel A	2	---
For the space between the shroud and the head stock f	---	6
For the thickness of the head stock f	---	12
Sum	4	6

	Ells	Inches
Sum carried forward	4	6
For the space to the wall e	1	---
For the thickness of the wall e	1	6
For the space between the sill g to the wall e	---	6
For the width of the sill g	---	14
For the space between the sill g and the pit wheel D	1	5
For the thickness of the pit wheel D	---	9
For the space between the pit wheel and the neck of the shaft	---	10
For the length of the neck	---	18
Sum for total length of the short shaft	10	2

The components on the long shaft B have the same thickness and width as those on the short shaft. Thus, to calculate the length of the long shaft, we just have to measure the distance from the inside face of the short shaft to the outermost side of the middle head stock and then add on the rest of the components on the shaft. Thus, subtracting the length of the neck of the short shaft, 18 inches, from 10 ells, 2 inches, the remainder is 9 ells, 8 inches for the length from the inside face to the outermost side of the middle head stock f, where it intersects the neck of the short shaft. Adding to this length 2 ells, 12 inches for the width of the channel and 1 ell 6 inches for the thickness of the outermost head stock and the neck of the shaft, the total length of the long shaft B is 13 ells, 2 inches.

Part 5

The third group of components has to do with the position of the shafts, which is based on the size of the gears in the vertical plane, as follows:

1) The two upper water wheels A A are 7 ells high and the pit wheels D D on each shaft have 64 cogs, spaced 4½ inches apart, on their periphery. The pitch radius[34] is 1 ell, 21 and 9/11 inches, and the radius to the periphery of the pit wheel is 2 ells, 2 and 9/11 inches.

2) The two lower water wheels are 7 ells, 12 inches high and have 8 blades per quarter. The associated pit wheels E E have 68 cogs, spaced 4½ inches apart. The pitch radius is 2 ells, 15/22 inches and the radius to the periphery of the pit wheel is 2 ells, 5 and 15/22 inches. The gear wheels generally have 7 staves and the millstones are 1 ell, 18 inches in diameter and 18 to 20 inches high.

3) With this gear ratio, the upper millstones rotate 9 and 1/7 times for one rotation of the water wheel. The lower millstones have the same dimensions as the upper ones, but for each rotation of the water wheel, they rotate 9 and 5/7 times.

Part 6

Now that we know the gear ratio, let's determine the position of the shafts:

	Ells	Inches
The space between the weir sill a **and the outermost edge of the water wheel A**	1	---
The radius of wheel A	3	12
Sum for the position of the first shaft from weir sill a	4	12

The following components are based on the position of the first shaft.

---

[34] Translator's note: "die Länge des Radezirkels bis zum Theilriß" The *Radezirkel* is the pitch circle of the cogged wheel. The *Theilriß* probably refers to the radius (as opposed to the diameter). Therefore, the length given seems to be the pitch radius.

	Ells	Inches
For the radius of the upper pit wheel D	2	2 and 9/11
For the space to the brayer beam h	---	5
For the thickness of the brayer beam h	---	4
For the overhang (Vorstich??) of the bridge tree i	---	6
For the space k between the bridge trees	---	18
For the Vorstich of the bridge trees	---	6
For the thickness of the brayer beam and the space to the face of the second pit wheel	---	9
For the radius of the second pit wheel	2	2 and 9/11
Sum	6	6

is the distance between the upper two shafts.

From the middle of the second shaft, the dimensions to determine the position of the fifth[35] shaft are as follows:

	Ells	Inches
For the radius of the second pit wheel	2	2 and 9/11
For the space to the brayer beam	---	5
For the thickness of the brayer beam	---	4
For the Vorstich of the bridge tree	---	6
For the space between the bridge trees	---	18
For the Vorstich of the bridge trees of the third mill	---	6
Sum	3	17 and 9/11

	Ells	Inches
Sum carried forward	3	17 and 9/11
For the thickness of the brayer beam and the space to the face of the third pit wheel	---	9
For the radius of the third pit wheel	2	5 and 15/22
Sum	6	8½

is the distance between the two middle shafts B and C.

---

[35] Translator's note: I think the author means *third* not *fifth*.

From the middle of the third shaft, the dimensions to determine the position of the fourth shaft are as follows:

	Ells	Inches
For the radius of the third pit wheel	2	5 and 15/22
For the space to the brayer beam	---	5
For the thickness of the brayer beam	---	4
For the Vorstich of the bridge tree	---	6
For the space between the bridge trees	---	18
For the Vorstich of the bridge trees of the fourth mill	---	6
For the thickness of the brayer beam and the space to the fourth pit wheel	---	9
For the radius of the fourth pit wheel	2	5 and 15/22
Sum	6	11 and 4/11

is the distance between the two lower shafts.

The fourth group of components relates to the length of the entire building as follows:

	Ells	Inches
For the thickness of the wall e	1	4
For the width of the solepiece l	---	16
For the space to the bridge tree	---	14
For the space to the pit wheel, including the aforementioned components	---	15
For the radius of the upper pit wheel	2	2 and 9/11
For the distance between the two upper shafts, from center to center	6	6
For the distance between the second and third shafts	6	8½
For the distance between the two lower shafts	6	11 and 4/11
For the radius of the fourth pit wheel	2	5 and 15/22
For the space between the face of the pit wheel and the last brayer beam	---	5
For the thickness of the brayer beam	---	4
For the Vorstich of the last two bridge trees	---	6
For the opening of the Wassertüre (water door??) F	1	18
For the thickness of the lower wall	1	4
Sum	30	2½

for the length of the building, including the walls.

The fifth group of components in the floor plan shown on Plate VI relates to the width of the building as follows:

	Ells	Inches
The thickness of the wall e	1	6
For the distance between the sill and the wall	---	6
For the width of the hursting	3	12
For the length of the bolting hutch G	3	6
For the space between the bolting hutch G and the partition H	3	6
For the width of the mill house to the wall	3	---
For the thickness of the wall	1	3
Sum	15	15

for the width of the building, including the walls.

Part 7

These are now the main components visible on the floor plan, Plate VI, on the horizontal plane. Plates VII and VIII show this mill from the vertical perspective. In particular, Plate VII represents the view from the water side and Plate VIII shows a cross section from the gable side. From this drawing, the height of the building, which involves the last important group of components, is calculated as follows:

	Ells	Inches
For the radius of the water wheel f g, measured from the cross-sill	3	12
For the radius of the pit wheel	2	2 and 9/11
For the space between the pit wheel D and the post of the hursting a	---	18
For the thickness of the posts	---	4
For the height of the millstones b	1	8
For the height of the hopper c from the top of the millstone	1	22
For the space between the hopper and the roof beams d	1	21
Sum	11	15 and 9/11

from the cross-sill of the upper water wheel to the roof beams d.

   Note

These are the most important components of the Staber mill with 4 millstones introduced in this chapter. The position of the other components can be seen in the various diagrams, whereby the dimensions can be determined from the scale bars.

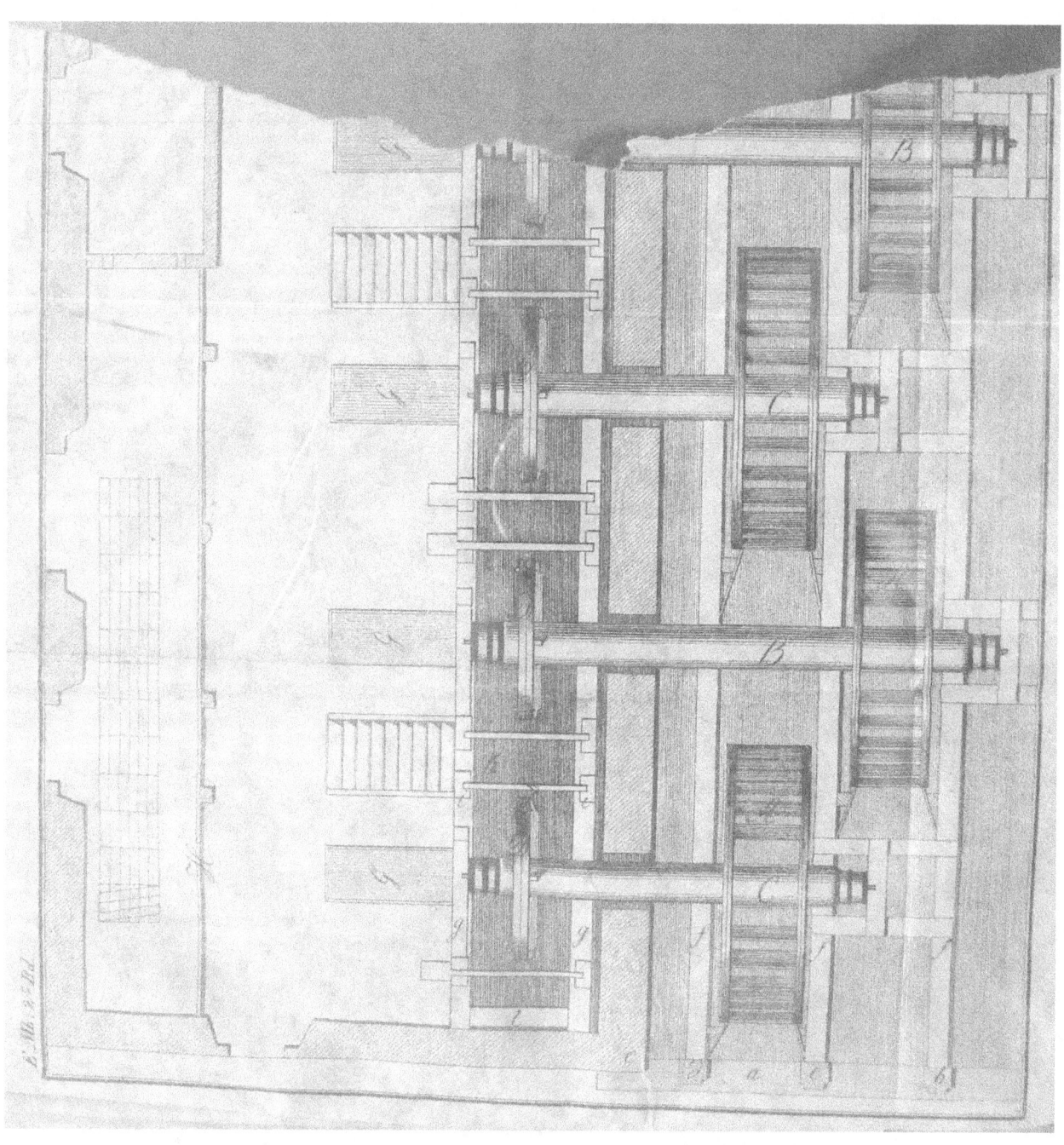

Plate VI

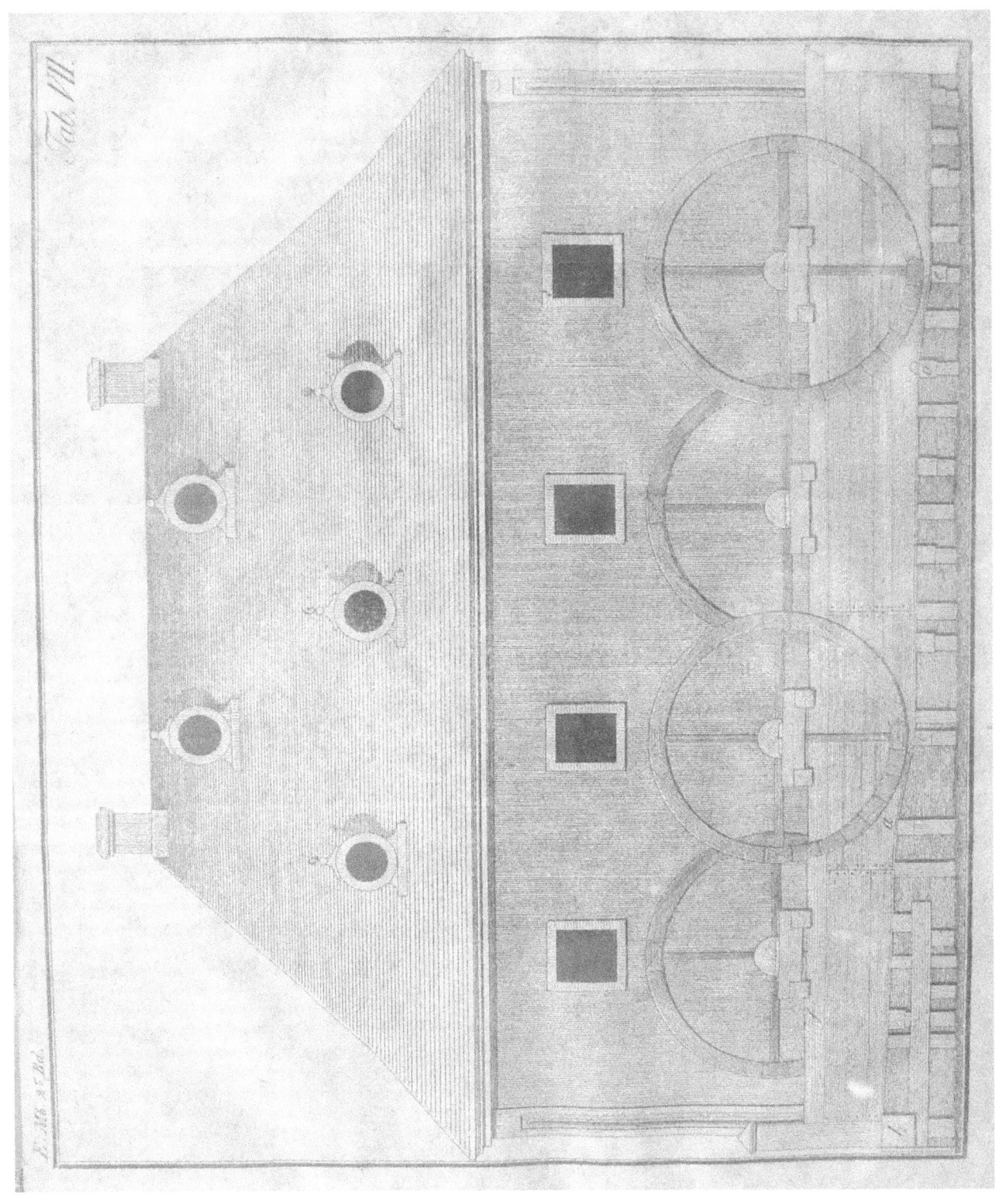

Plate VII

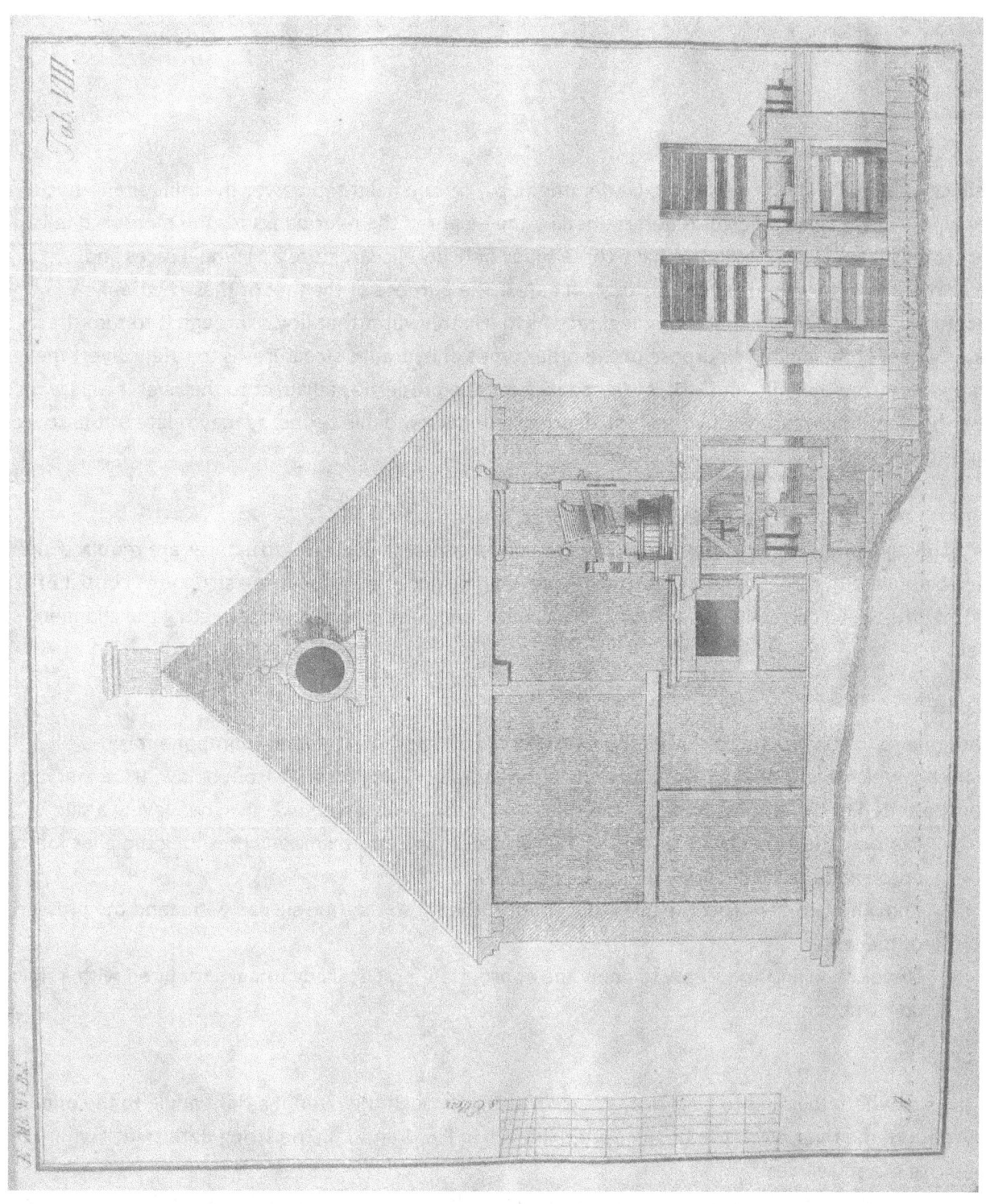

Plate VIII

Chapter 4

Sluices

Part 1

A sluice for a mill allows the water to be dammed up a certain height to power the mill (or another machine). The allowable height is determined by the height of the river banks for the measured fall and the established mill laws. Mills can have three types of hydraulic structures: 1) wheel races and spillways, 2) overflow weirs and 3) vertical lift gates. The purpose of the first of these hydraulic structures, namely, the mill race or wheel race, is to use the water that flows through it to turn the water wheel of the mill. The purpose of the other types of hydraulic structures is to safely divert the extra water from the mills and also, so to speak, from the properties adjacent to the river. I will try to describe these three types of hydraulic structures in understandable terms, as they relate to the scope of this book.

Part 2

For a sluice for an underflow water mill, for example, a panster mill, three structures are required and must be built with care if the sluice is to work well and be durable. These three structures consist of: 1) building the apron correctly, 2) stabilizing the weir sill, and 3) appropriately connecting the channel itself.

Part 3

With regard to constructing a sluice for a panster mill, for example, the main component is the sill *a* (Plate IX), which is responsible for preventing water from breaking through from below. Its preservation must be suited to the ground. Here are the rules that must be observed exactly when laying a sill:

1) Examine the ground with an earth auger to determine approximately how long the piles for the apron and the foundation need to be.
2) Know how to construct the piles that support the sill so that the sill can withstand the pressure of the water.
3) Understand and know how to apply the appropriate practical advantages required with a sluice construction.

Part 4

At the site where the sill is to be laid, 1) stretch a string through the river at a right angle to the middle flow path of the river, as shown by the dotted line a b in Fig. 1 and 3[36]. The string determines where the middle of the sill will rest.

2) Depending on the length of the sill, double piles may be driven, as shown in Fig. 3. Contrary to Beier and Melzer's instructions, and those of others referenced in their mill construction books, do not drive the piles next to each other, but rather stagger them a width apart as shown by piles c, d, e, and f in Fig. 3. When the piles are staggered, there is not as much pressure on the tenon axle as when the two piles are driven against each other, where the small amount of wood between the double tenons is

---

[36] Translator's note: I think the author is referring to Plate IX, although there is no dotted a b line in Fig. 1.

easily weakened. Instead, the sill becomes more stable. The distance between the piles and the length of the piles depends on the nature of the river bottom. Thus if the ground is loose, then it is understood that the piles must be driven closer together than when the ground is firmer.

The foundation piles for the sill are usually 5 to 6 feet apart in moderately firm ground. They are 10 to 14 feet long, depending on the nature of the river bottom.

3) Once the foundation piles for the sill c, d, e, f and so on have been properly driven, the position of the sill will be determined by the fall height, as allowed by the mill law and which has been measured. The procedure is as follows: Based on the previously mentioned average normal water depth, stake out the fall. The uppermost section determines the position of the top of the sill. If, however, a mill is already standing there, then it is understood that the position of the new sill must, of course, be determined by the depth gauge[37].

4) After you note the exact position of the sill, mark the thickness of the sill from this point down the depth gauge. The latter point shows where the piles c, d, e, and f in Fig. 3 must be chiseled in; make sure, however, that you add 6 to 8 inches to the tenon on the pile from this point upwards.

5) Hold a straight edge against the lower edge of the sill. The straight edge must be long enough to reach from one pile to another. Place a proper plumb-bob level on the straight edge and level the shoulders (or the holes for the tenons) on all piles upon which the sill rests. Write down the width of the tenons and cut them. Then cut the mortises based on the position of the corresponding piles. After the carpentry is done properly, the sill is ready to be installed. The shoulders of the sub-sill beams g g in Fig. 3 (previously mentioned in Chapter 2, Section 17) – specifically the lower side of the plates h h where they connect underneath the sill – lie 6 inches below the shoulders of the foundation piles c, d, e, and f. However, the tenons on these beams, which go through the sill, must be longer than the tenons from the foundation piles by the thickness of the plate h from the sub-sill g. Install the sill a in Fig. 1 so that it lies in the horizontal plane along both its width and its length. Finally

6)[38] Stabilize the sill with piles as follows: Construct 4 inch thick planks out of pine; they can be 18 to 24 inches wide. Their length, however, depends on the nature of the river bottom, as was the case with the other piles. For moderately solid ground, they are usually 5 to 6 ells long; if the ground is loose, however, the piles should be made 10 to 12 ells long. The sill planks are one of the most important things when building a sluice. For this reason, they must be made of good, strong wood and so precisely worked that all joints are full and not splintered. Sometimes the grooves are cut in the joints of the sill planks, which have feathers installed on the other part. This device alone is only required for sluices built in rivers with quicksand or other types of sandy bottoms. Otherwise straight sill planks, tightly joined, are preferred. Other things pertaining to the construction of the sill planks can be seen clearly in

---

[37] Translator's note: The *Mahlpfahl* or *Sicherpfahl* was usually an oak post driven into the river bottom. It shows the legal water depth and the height of the weir sill. It was used to prevent millers from damming water higher than the level allowed by the mill law. Source: http://www.zeno.org/Pierer-1857/A/Aichpfahl and http://wapedia.mobi/de/Sicherpfahl.

[38] Translator's note: Mistakenly numbered "5" in the German original.

Fig. 2. In this figure, some of the sill planks are shown as they would stand in the bottom of the river. The planks are sharpened only on one side, to force the joints closer together, and that is how the sill will be stabilized.

1) Note

When constructing a sluice where the river bottom is favorable and consists of soils that hold the water, for example, clay loams or clays, then it is not necessary to stabilize the sill with planks. Instead, a wall is put behind the sill, which is made of tongue-and-groove joined posts, 3 inches thick. These posts can be driven 5 to 6 feet deep into the ground, measured from the surface of the sill. Strong, pointed strips can be nailed across the lower posts; these strips project over the ends of the posts by 2 to 3 feet, allowing them to be driven into the ground. If you want these posts to have watertight joints, then use a circular plane to make a groove in the joint. Lay a wick made of oakum treated with tar into the groove, which prevents even one drop of water from penetrating. This method is also recommended for overshot water channels; more about this where the construction of overshot water mills is described.

Part 5

After the sill has been laid and properly stabilized, as described in the last section, we turn our attention to constructing the apron (Plate IX, Fig. O) as follows:

1) Starting from the weir sill a in Fig. 1, three to four sills b, c, d, and e, are joined transversely across the river by mortise and tenon to foundation piles. These piles can be 4 to 5 feet apart. These piles are chiseled out in such a way that any given sill lies 8 to 12 inches deeper, depending on how deep the river bottom is, than the previous one. For example, the first sill b can be 8 inches deeper than the weir sill a, the second sill c just as much and the same with the others. Proceeding in this manner, the fourth sill (a main sill e different from the weir sill) will lie 1 ell, 8 inches lower than the surface of the weir sill a. Consequently, the apron is sloped away from the weir sill. The inclined apron is necessary not only to prevent silt from collecting on the apron, but also to prevent breakthrough from below. When the river bottom is watertight, however, the apron does not really have to be inclined, because under these conditions, the ground already protects against breakthrough from below. After
2) these sills have been positioned correctly, the first sill e is again stabilized with planking. I want to make the following comments about driving the sill planks. When you begin driving the piles into the ground, don't start at the end of the sill, like some do, but rather in the middle. This is why the first sill plank has to be sharpened on two sides whereas the others only have to be sharpened on one side, as previously mentioned. The reason for this was explained in Part 4. If you have weak sill planks, however, drive two rows of planks so that the second row exactly overlaps the joints in the first. After stabilizing the first main sill e with planking, now
3) drive the piles for the side walls, and chisel out the head stocks ff in Fig. O, depending on the height of the bank. With this, the framework is finished and ready to be lined. Before starting the lining process, carefully
4) tamp clay or another soil that holds water into the spaces between the mudsills c, d, and e. While doing so, make sure that only a very thin layer of loose soil or clay is put on the tamped down ground, otherwise the tamping will not hold. Continue this process of tamping until the

level is the same as that of the sills, and then add about another quarter inch of loose soil on top. Now this structure is ready to be lined with planks. With regard to

5) lining the apron with planks, if everything is to be watertight and stable, I have to mention the following. Assuming that all planks are very precisely joined, before you lay the planks, coat the heads of the planks with hot tar, which is mixed with pitch. After doing so, treat them with fine, fibrous aquatic moss and put tar on both ends of the planks. Now lay them next to each other and nail them down with the well-known plank nails, as shown by *g* in Fig. O. Before nailing down the second plank, however, use a rounded tamper to tamp pure soil underneath the first nailed-down plank so that the soil is uniformly even and closely packed under the plank. In this fashion, continue to nail down the planks until the entire apron is covered. The weir sill *a* gets a groove *h*, which is 4 inches wide and as deep as the thickness of the piles allows. This is necessary because the mill laws do not allow the sills to be lined with planks.

2) Note

Covering the side walls f f with planks requires as much care as lining the bottom of the channel, as mentioned previously in Part 5. One small opening, which is hardly noticeable, through which water just seeps at first, can later become so bad that stopping the leak requires the entire side wall to be torn out. This is what I recently personally experienced at a panster mill, where even the planks stabilizing the weir sill, which stood in the bank, were washed out. Wouldn't it also be possible for such a breakthrough to wash out all the soil underneath the apron? For this reason, all old wood and rocks must be carefully removed from behind the side walls until you reach solid ground. Then pure soil must also be added on top and each layer must be tamped down. The blue light soil doesn't just conserve the posts for a long time, it also hinders breakthrough when a piece of post has rotted.

Part 6

Now that we've described how to build the apron and how to stabilize the sill, we turn our attention to the lower part of this panster sluice. The lower part consists of a wheel race and a spillway. The wheel race is constructed as follows:

1) The piles needed for the head stocks *k k* are driven in for however long the channel is to be, in a straight line starting from the middle of the two sluice posts *i i* in Fig. 1. The distance between these piles depends on the length of the beams. Usually you use 8-ell long beams and you figure 4 piles for one beam. Based on this calculation, the piles should stand 2 ells, 16 inches apart. Now, on these piles

2[39]) the head stocks *k k* are mortised so that when the first panster water wheel is hung in its lowest position, the surface of the head stocks is still about 2 inches from the shaft of the water wheel. Hook tenons on the ends of the head stocks are inserted into mortises in the sluice posts and joined in this way. When you are finished with

3) the head stocks *k k*, the cross trees *l l* are leveled according to the incoming fall height and joined by mortise and tenon to the piles at the appropriate position, depending on whether this

---

[39] Translator's note: Mistakenly numbered "3" in the German original.

is to become a flat runway or an apron. More will be said on this topic under the construction of panster mills. Now, lay

4) the sub-sills *n n* in Fig. 1 and the remaining mudsills *m m m* and so on 2 ells apart from each other and then line them with planks, designated *o* in Fig. 1. Now the wheel race is finished.

Part 7

The design of the spillway can easily be seen in the drawing. However, the main unit must be laid out in such a way that the water is carried away through the spillway 1) without "boiling" in front of the channel and 2) when there is ice on the river, that the ice floes break on the spillway. To ensure that the spillway possesses these two characteristics, proceed as follows: Lay the middle mudsill m in Fig. ♀ in Fig. 1, rather than the last sill, deep enough for the measured fall. Then let the bottom of the channel slope 8 to 10 inches according to the level. In this way, the aforementioned characteristics can be achieved.

The length of the spillway for panster mills is usually 8 to 9-fold the normal water depth ahead of the sluice gate. For example, if the water depth were 1 ell, 12 inches, then the length of the spillway would have to be 12 ells. There is no harm in making the spillway 3 or 4 ells longer than the aforementioned rule of thumb suggests, however. During times of high water, the longer the spillway, the better the unusable water level is thrown out and the length helps more water to be removed from the wheel race.

Part 8

We have just described the layout of a sluice for a panster mill as it pertains to the scope of this book. Now we need to describe the designation and position of the parts in the drawing, which were not mentioned in the previous section. I will also give practical advice to my readers who are still learning about this subject.

Fig. 1

a	the weir sill
b c d e	the mudsills of the apron
ff	the head stocks of the side walls of the apron
g	a piece of the apron lined with planks
o	the depth gauge
h	the stabilizing planks of the sill
i i etc.	the sluice posts
k k	the head stocks of the wheel race and the spillway; the spillway is shown in Fig. ♀
l l	the cross-trees
m m etc.	the mudsills of the wheel race and the spillway
n n	the sub-sills
p p	the side beam sills
q q	the side beams

Fig. 2 shows the side view of the sluice, with the following components:

a   part of the stabilizing planks of the sill
b   the sill
c   the panster gate
d   the bracing
e   the panster safety shaft (Schutzwelle??) with chains

Fig. 3 shows the ground plan of the sill and how it is attached to the piles. Part 4 explains the details clearly enough. The other components can be seen sufficiently well in the drawing and their dimensions can be determined from the scale bar.

Part 9

I cannot neglect to tell my practitioner-readers a few rules of thumb with regard to working and positioning the different types of posts that are needed for mill sluices. These rules pertain to: 1) the naming of the posts and 2) the carpentry work involved.

Part 10

Pointed posts are those that are sharpened on three or four sides.

These pointed posts can, in turn, be divided into two groups: a) ground posts and b) land posts. Ground posts either completely stand in the ground (such as, for example, the piles) or in the ground and in the water at the same time, such as, for example, the foundation piles under the undershot channels. On the other hand, land posts stand partially in and partially over the ground and the water, such as, for example, the head stocks and the side walls of the mill races. The second type of post is the so-called stabilizing planks or sheet pilings. These are sharpened on only two sides and possess either a flat joint or a groove and feather, depending on the river bottom, as previously described in Part 4.

Part 11

As mentioned previously, the pointed posts are sharpened on either three or four sides. In both cases, the length of the tip is three times the thickness of the post. Regarding the difference between the three- and four-sided pointed posts, the former is preferred to the latter, because they are easier to drive into the ground. Furthermore, they don't twist and turn as easily as four-sided and round tips. The sides of the tip should not come together into a point; rather the tip should be flattened, as shown in Fig. 1 of Plate X.

Part 12

The tips of the posts are usually cut from the weaker (toward the treetop) end of the trunk, but in most cases it's better to make them from the strong, lower end of the trunk. If it is not necessary to cut them, it is better to drive the posts as they are when they come out of the woods. If no heavy load is placed on the posts, such as, for example, with head stocks, sluice walls, and similar structures, then, if they are driven in with the weaker end, they will often be lifted out by rising water because of the ice that has frozen onto them. I know this from personal experience and could name a number of examples. In the second case, however, where the strong end is driven in, it is less likely that this misfortune will occur.

The aforementioned example is not the only reason why posts are pulled out of the ground. There are several other causes, for example, springs[40] that have a large upward force, compact and firm soil, when other posts are driven in next to a given post. During thaw, frost also lifts posts and even entire weirs along with their sills out of the ground, etc. These examples show clearly that it is better to drive posts that do not have to bear a heavy load into the ground with their strong end in order to prevent this kind of misfortune.

We know from experience that driving posts into the ground with their strong end is somewhat difficult at first, but gets easier after that. It takes about one-fourth more time than the usual method in which the weaker end is driven into the ground. Doesn't it make more sense to spend a little more time and money now than to have a hydraulic structure, built with a lot of effort and money, ruined in a few years? I want to point out one more thing with regard to driving in the strong end: The weaker end (toward the treetop), may not be too weak, because the head of the post could easily split while driving it in or it could develop whiskers. The consequence of the latter misfortune is that you have to drive it in longer.

Part 13

The length of the piles depends on the nature of the river bottom, as already mentioned in Part 4. For this reason, I can't say anything specific about the length of the piles in a planned hydraulic structure. To approximate this length, we could test the river bottom using an earth auger, as mentioned in Part 3, or by driving in a test pile. These methods are not completely reliable, however, and you should not trust the most likely probability. Rather, make the piles a little longer than what the aforementioned test says. It is often the case, when the soil is poor, that even very long trees used for piles are not long enough to reach solid ground when it is covered with muddy soil types. Under these conditions, these piles must be connected end to end, or spliced together. Perronet, a famous French architect, mentions a method for splicing piles end to end, which is shown in Plate X, Fig. 2. This method is quite suitable, but just make sure that the lock is worked very precisely and, furthermore, that it is surrounded by strong rings. I have known about this method for a long time, since I once used it to make a Staber water wheel shaft from two pieces. Not only did I have the entire lock protected from coming apart by using good, waterproof putty, I had it surrounded by three rings. This spliced shaft, which consisted of a new and an old piece, worked for more than six years before it had to be replaced.

Part 14

The load that a pile, which is driven in to maximum stability, is to bear depends a lot on the compressed soil against its thickness and length, but also on the strength of the wood itself. As yet, no meaningful experiments have been carried out to determine the strength of a pile, taking its length into consideration, when subjected to a given load. Scholars have come up with a rule of thumb calculation for estimating the strength of vertically-standing piles for the load to be borne. The strength is proportional to the square of the thickness of the pile, where the cross section makes a right angle to the pile, multiplied by the width and divided by the square of the length of the pile. In this case, the

---

[40] Translator's note: In this sense, spring means water from the ground.

thickness of the pile is understood to be the narrow side of the horizontal cross section. The length is understood to be the height of the pile above the ground. Thus if we have two piles with the same cross-sectional area, for example, each one 6 square inches, and if one stands 6 feet and the other 12 feet above the ground, then according to this calculation, the shorter pile (6 feet long) should be able to carry almost four times the load of the longer pile (12 feet long). However, we cannot use this calculation in practice all the time, because we cannot always choose the thickness of the piles we use in a hydraulic structure. That is why it is better to drive the piles a little closer together, so that they follow the generally accepted practical rule of thumb of spacing them 4 to 4½ feet apart. With this method, it is also possible to make use of weaker piles, although they still have to be driven in far enough to attain the proper stability.

Part 15

The sheet piles mentioned in Part 10 are worked differently depending on the firmness of the ground. If the ground consists of quicksand or another loose soil, then the kind of sheet piles shown in Fig. 3 of Plate X are best. In these piles, the feathers and grooves are worked at right angles with the sides of the piles. The recess of the pile and the elevation of the feather are 3 to 4 inches. In the event that there is a piece of wood that gives two pile lengths, then grooves are routinely made on both sides, as shown in Fig. 4. The adjacent pile then gets two feathers, whereby the weaker end of the wood can be chosen for the latter. When the ground is firm, rather than soft, stronger piles must always be chosen for the sheet piles or the piles that stabilize the sill so that they do not bend when driven in. For soft ground, the piles can be 4 to 5 inches thick; for firm ground, they should be 6 to 7 inches thick. With 4 to 5 inch-thick sheet piles, it's best to carve a shingle groove, as shown in Fig. 4 on Plate X. The advantage of this type of joint is that the side wood can't break off as easily as it would with weaker sheet piles that have a groove at right angles. When the ground is rocky, the pointed posts and the sheet piles must be covered with iron shoes. The shoes are attached to the pointed posts with three or four feathers, depending on whether the posts are sharpened on three or four sides. The feathers are nailed onto the posts. Fig. 5 shows how the shoes are attached to the pointed posts and Fig. 6 shows how they are attached to the sheet piles.[41] The tips of the piles are not worked in the same way as those without shoes; rather the tips should be worked to stand flat in the shoes. Furthermore, make elongated nail holes in the feathers so that the shoe can "give" when the wood is compressed slightly on the iron surface inside the shoe when the post is driven in. Otherwise, the nails will be pulled out during this process.

3) Note

The practical advice and rules of thumb that I recommend you follow when constructing sluices also apply to all similar hydraulic structures. The one exception, where there are other advantages to note, is a similar hydraulic structure made of stones, which will be described elsewhere. Other than that, I do not think that it is necessary to repeat the aforementioned rules and regulations in the chapter on weir construction; rather I will only mention those required in addition.

---

[41] Translator's note: I don't see any feather-shoe detail in Fig. 5 and 6.

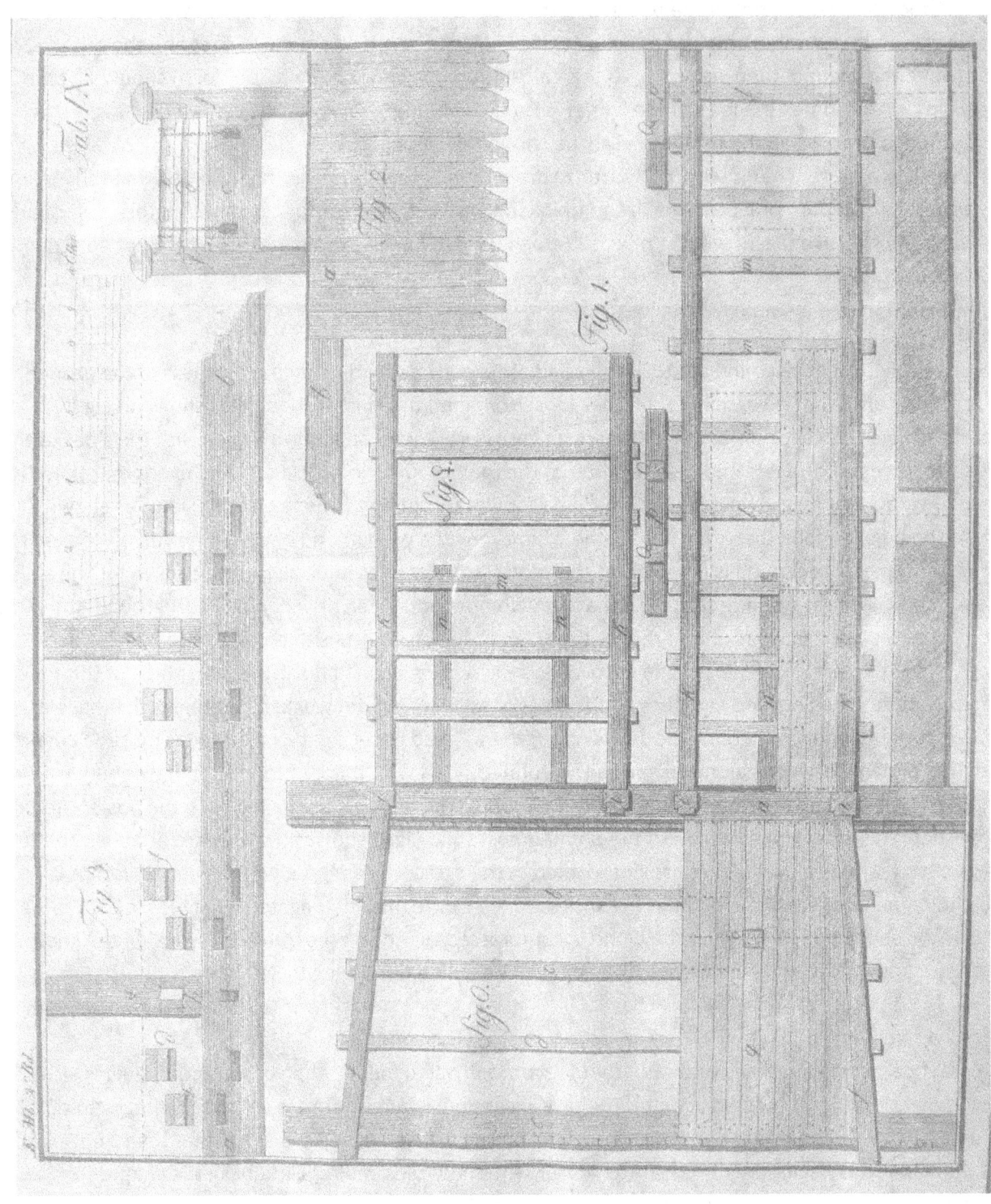

Plate IX: Figures 1 through 3

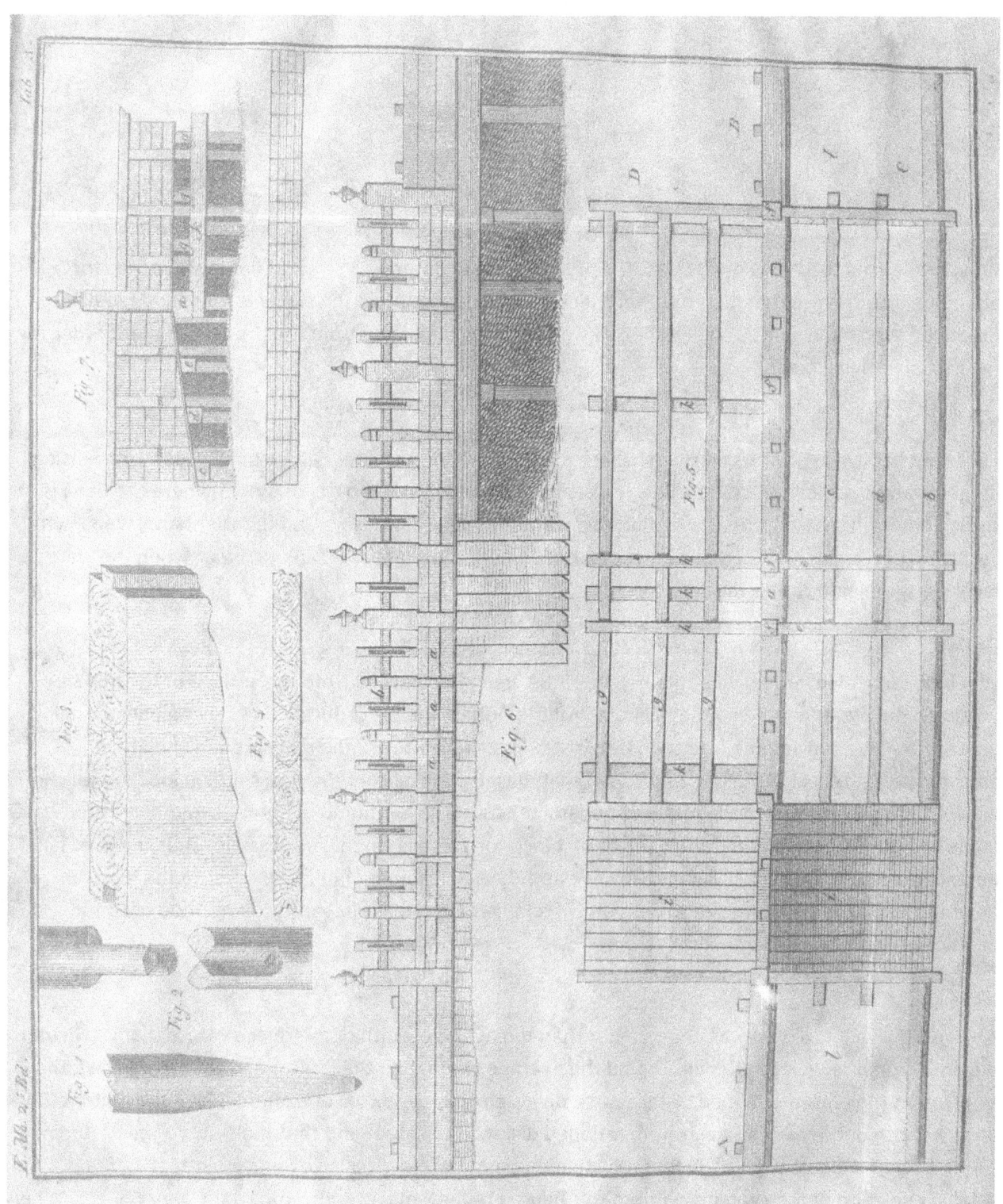

Plate X: Figures 1 through 7

Chapter 5

Weirs

Part 1

Why a weir is needed for a mill was stated previously in Chapter 1. Weirs and dams are means for damming up a river so that mills or other mechanical devices obtain the water level needed to drive the machine. Weirs and dams also divert extra water from the mill and from the properties that border the river to lower lying areas.

Part 2

There are two kinds of weirs on rivers where mills and other machines can be built: those with vertical lift gates and overflow weirs. Overflow weirs are only used in those parts of a river where the bank is sufficiently high that we can be sure flooding won't happen very easily. On the other hand, weirs with vertical lift gates are built in those parts of a river where there are low banks, so that when the water level rises, the gate can be raised.

Part 3

When building an overflow weir, the height of the weir sill is leveled from the surface of the mill sill based on the water depth and the fall height of the outlet. The procedure follows. For a panster mill, let's say that we can assume a water depth of 1 ell, 12 inches above the measured fall height, in accordance with the mill laws. Let's also say that, based on local conditions, a suitable location for the weir is 1200 ells away from the mill. The general practical rule of thumb in the millwright trade for water inlets and outlets is to provide a fall height of 1 inch for 100 ells whenever possible. Thus the inlet fall height comes to 1200 ells, 12 inches. Now we add the water depth (1 ell, 12 inches) to the inlet fall height to get 2 ells as the height of the weir sill relative to the mill sill. This is the method used to position weirs relative to the sills.

Part 4

An overflow weir is always one-third longer than the average width of the river at a site that is twice the distance from the weir to the mill. To find the average width, find the largest and the smallest width of the river at the aforementioned distance and divide the sum by two. For example, let's say that the largest width of the river at the aforementioned distance is 50 feet and the smallest is 34 feet. Adding the two widths together gives 84 feet. Dividing this sum by two gives an average width of 42 feet. Adding one-third of the average width to this figure (14 feet) makes the width of the weir 56 feet. When building a weir with a vertical lift gate, however, the width of the weir always has to be the same as the average width of the river, including the width of the sluice posts and set piles. This kind of weir can, in principle, be narrower, because the weir sill is laid at a position which is not the normal water depth plus the inlet fall height, but rather only at a position that represents the fall height relative to the mill sill.

Part 5

Plate X, Fig. 5, 6, and 7 show the layout of the components of a weir with a vertical lift gate. When building the weir:

1) Follow the rules and regulations described in Chapter 4, Part 3 when driving the piles for the weir sill *a* in Fig. 5.
2) Completely cover the weir sill *a* and the last mudsill *b* with planks, as described in the previous chapter. Taking the nature of the ground into consideration,
3) to stabilize the apron, lay two sills *c* and *d*, or more, depending on whether the apron is to be long or short, between the weir sill *a* and sill *b*. To assemble the rest of the components, follow the directions given for panster sluices.

Part 6

The design of the rest of the weir with a vertical lift gate described here can be seen clearly in the drawing. The dimensions of the components can be determined from the scale bar. Thus I just need to give a short explanation of the views along with the components. Fig. 5 is the plan view. Fig. 6 is the cross-sectional view from the perspective of the long side of the sill *A B*. Fig. 7 is the side view. Fig. 5 shows the following components on a horizontal plane:

a   the sill

b c d   the apron sills

e e   the rear head stocks

f f etc. the sluice posts

g g   the front sills of the spillway

h h   the front head stocks

i i   a section of the floorboards with posts

k k etc. the sub-sills

l l   the chambers inside the banks, which may be framed

Fig. 6 shows the entire sluice frame in the vertical plane. Note that the gate *a a* has slotted arms, so that the gate can be easily lifted by means of a lever that has an iron rail. The lever is inserted through the slot with the rail, with its short end on top of the horizontal brace *b*. Pulling up on the long end of the lever causes the rail to press against the pin, lifting up the gate.

Fig. 7, the side view from the perspective of *C D* in the plan view, shows the following components:

k	the outermost sub-sill
c d e	the apron sills viewed from the end
a	the weir sill
m m etc.	the piles for the side walls along with the other components that are apparent from looking at Fig. 7.

Part 7

How to determine the height of an overflow weir relative to the position of the mill sill was explained in Part 3. Thus we only have to concern ourselves with the steps required to build such a hydraulic structure, as far as it pertains to the scope of this book. This topic is covered in greater detail than is possible here in all those books that deal directly with hydraulic structures.

Part 8

Accordingly, Plate XI shows an overflow weir made of stones or a dam. The steps for constructing it are as follows:

1) After directing laborers to dig out 3 to 4 ells below the normal river bottom *A* in Fig. 2, lay
2) the pilotage, or the foundation piles for such a hydraulic structure. These piles
3) are laid out as follows: Drive foundation piles into the ground 3 ells apart for the length of the dam *A B*. Then
4) lay the sills *a, b, c, d,* etc. on top of these foundation piles by means of precisely carved mortise and tenon joints. Lay the solepieces *e, f, g, h,* etc. on top of these sills, so that they are cogged transversely onto the sills *a, b, c, d,* etc.
5[42]) The beams form a grid, as seen clearly in the drawing, with individual compartments (or fields) *i, k, l,* and *m*. After finishing the pilotage,
6) add a layer of loose soil and level it out. Then add a layer of field stone or rubble, as shown in Fig. 2. Then tamp down gravel on top of these stones and embed the stone pavers.
7) Cover the two sills *a* and *b* in Fig. 2 with planks, and also stabilize the inner side at *c* with strong posts. All else can be seen more clearly in the drawing.

Part 9

The main thing about constructing an overflow weir or dam is to get the right curvature. This curvature involves the line of an arc, such that the water not only flows smoothly over the dam crest, but also gradually approaches the horizontal direction below the dam at point *d* without generating a roiling wave. The best curvature, proven by experience with well-designed dams, is explained here, and is shown in Fig. 2:

---

[42] Translator's note: This item was mistakenly numbered "4," making the subsequent numbered items off by one.

1) Make a straight line *e d* from the endpoint of the lowest sill *b* to the endpoint of the uppermost sill *c*.
2) Divide this line *d e* into 5 equal parts and mark the third part at *f*. Furthermore, also divide
3) the width of the dam, namely from *g* to *h*, into 5 parts, and use the 3 parts from *h* to *i* to set the compass aperture.
4) Using this compass aperture, strike intersecting circles centered at *e* and *f*, which intersect at 1^{43}. Place the center of the compass at this intersection 1 and swing the upper arc *e f*.
5) Take the distance from the outermost point of the lowest sill *b* to point *i*, and strike intersecting circles centered at *k* and *f*. Then place the compass at this intersection and swing the lower arc *k f* with this aperture.

1) Note

The lowest sill *b* can be laid 6 to 8 inches below the usual water depth *d* so that the water below the dam always stands in the arc *k f*. In this way, the damaging wave that results from water falling over the dam by the force of gravity is not only dispersed, but the water also always flows away from the lower part of the dam in a straight line.

2) Note

To really protect the lower part of the dam from being washed out by the water, build a fascine d^{44}, laying the solepieces *l* on piles and on the sill *k*. This construction can be seen in the drawing.

Part 10

When building a weir from stones, we still have to answer the following question: How wide does this kind of weir (or dam) have to be to completely withstand the force of the water lying ahead of it? The field of hydrostatics provides a rule of thumb with which we can calculate the side pressure of the water against a vertical surface. However, this calculation is more applicable to a vertical surface against which water presses directly, not specifically to an earth and stone mass which makes up the weir. Because scholars do not yet agree on all these calculations so that they can actually be used in practice, we have come up with another practical way to approximate how thick or wide a stone weir has to be to be able to withstand the head of the dammed up water. To do so, look for the correct average water depth over a stretch of 12 to 1400 ells of the river at the site where the weir is to be built. Multiply this depth by 5 to get the width of the weir. For example, let's say that we find an average water depth of 3 ells. Multiplying this depth by 5 gives 15 ells as the width of the weir. This method was used to determine the width of the majority of stone weirs that have been built. They have adequately withstood the pressure of the dammed up water.

Part 11

In addition, I recommend that the following practical rules of thumb be carefully observed when building a stone weir:

---

[43] Translator's note: See small x with the number 1 below it at the top of Fig. 1.
[44] Translator's note: The author probably means upper case D. Technical annotation: Shown in rubble at lower base of the dam.

1) It is always better to make these kinds of hydraulic structures, which weirs are, a little stronger than the aforementioned rule suggests, so that they can hold back the dammed up water and the ground above them.
2) The construction materials, wood and stones as well as lime, required to build a weir must be selected carefully and handled properly when building the structure. The rubble stones used to cover a dam must not be too small. They have to be at least 1 ell high, otherwise the weir will never be strong enough.
3) The pavers that line the surface of the dam must be well-covered with moss. The advantage of having a moss-covered dam is that when ice and log rafts go over the dam, the pavers are not knocked loose, as they are on dams not covered with moss. This is the misfortune that causes stone dams to develop cracks.
4) Before beginning construction of such a hydraulic structure, you must also pay attention to the time of year when floods are not likely to be a problem. The best time is often in the spring after Easter. However, I'm not suggesting this as a general rule, but rather a precautionary rule of thumb, because floods can occur any time. Regardless, it is still better to begin the construction of such a structure earlier rather than later in the year.
5) With a solid hydraulic structure, a master builder must mainly be vigilant that the pavers for a dam are well leveled by the bricklayers or stonemasons on their adjacent sides so that little stones cannot wedge themselves in the cracks. This is a bad situation, because the wedge-shaped form of such a stone makes it susceptible to being pulled out, and one can never be sure that the dam will not develop a crack.

3) Note

I am well aware that there are still many other obstacles that one encounters when building a hydraulic structure, which often cannot be anticipated. Thus here I have only written as much about these hydraulic structures as pertains to the scope of this book. If God gives me a long life and good health, however, I plan to write a special work with more detail about hydraulic structures for mills for my honorable readers, in which I will also describe the machines required for building these structures.

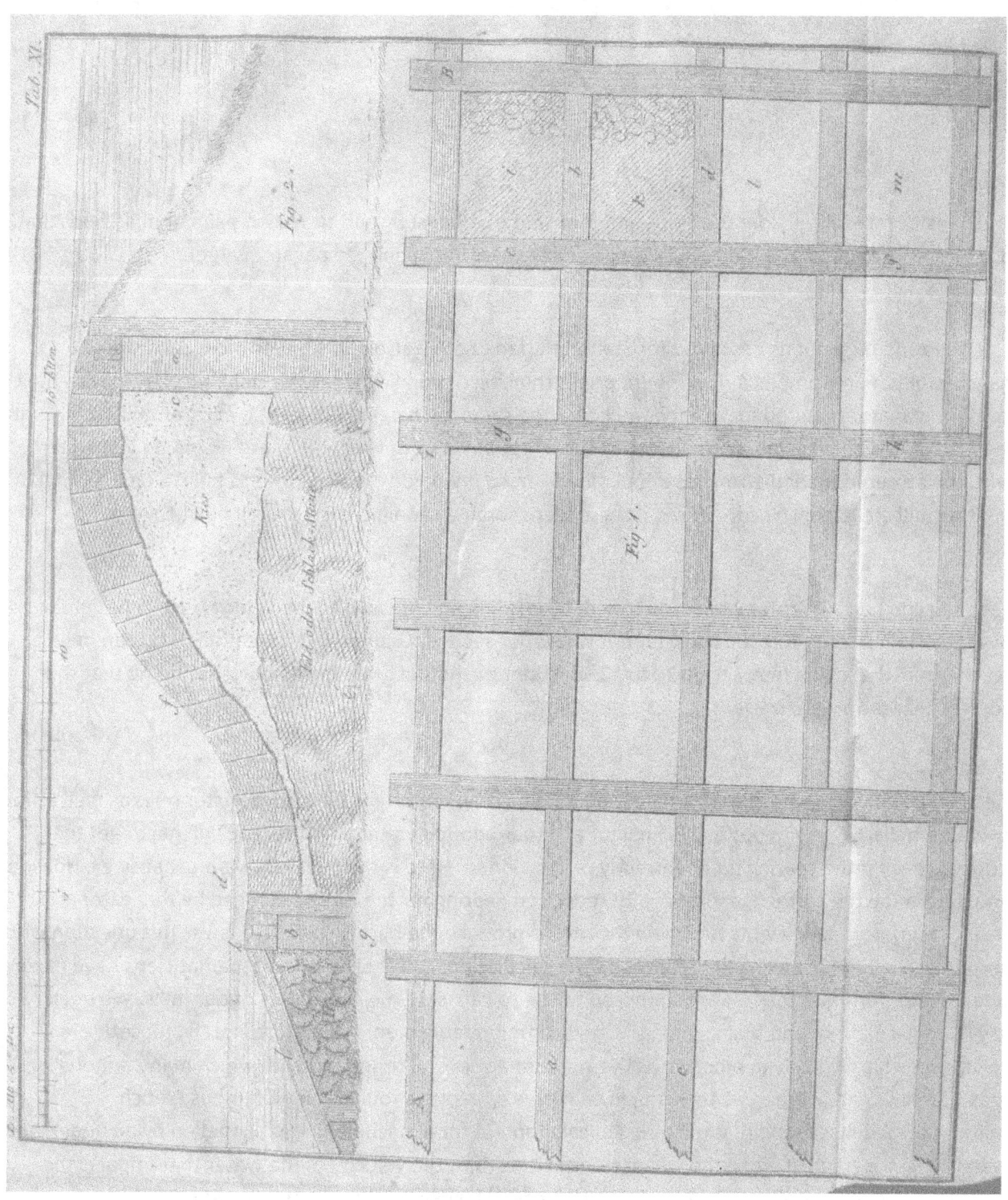

Plate XI: Figures 1 and 2

Chapter 6

Panster Mills

Part 1

The panster mills differ from the Staber mills only in that panster mills have two pairs of millstones that produce an increased effect due to twice as much water impacting the panster wheel.

Part 2

The rules and regulations described for Staber mills in Chap. 2 also apply under most conditions to panster mills. For example, a river where one is thinking of building a panster mill must have twice the amount of water required for a Staber mill over the same fall height. Similarly, a panster mill requires at least 10 inches of lively fall. With 90 to 100 cubic feet of water per second, there would be less water with the aforementioned 10 inches of fall, however, so it wouldn't make sense to build a panster mill. A panster mill could only be considered if there were some way to increase the fall height.

Part 3

Let's say you had a river for which you found, by leveling and measuring the velocity, 10 inches of fall over a certain stretch and 100 cubic feet of water per second. Under these conditions, you can build a panster mill that could finely grind 10 to 12 Dresdner bushels of grain in 24 hours, assuming that the correct gear ratio were present.

Part 4

Getting the correct gear ratio with reduction gearing is an important thing for panster mills as well as for overshot mills. Even with optimal fall height and the required water quantity, you will never get the desired result if the gearing is not correctly proportioned. Here is the question we must answer: How do you determine the correct gear ratio with reduction gearing for the force produced by the water as a result of different fall heights to enable the mill to produce the best result? To answer this question with certainty, the field of mechanics teaches us the calculations to use. Famous mathematicians have given us analytical formulas that we are supposed to use to calculate the gear ratios of our mills. We need to say "Congratulations and thank you!" to these commendable men, but, unfortunately, these calculations hardly apply in practice, as I've mentioned several times and with which many famous master millwrights will agree. An example will serve as explanation. Parent, a famous French mathematician, has found that the mechanical moment for a machine that is propelled by an undershot water wheel is greatest when the wheel turns at one-third the velocity of the water that impacts the wheel. For example, if a panster wheel had 10 inches of lively fall, this wheel should be 8 ells high, according to our practical rule of thumb, when the diameter of the millstones is 1 ell, 12 to 13 inches. With the correct gear ratio and the mill running at full speed, the water wheel would complete a revolution in 7 seconds. According to Parent's calculation, however, the water wheel requires $22\frac{1}{3}$ seconds for one revolution. Think about how slowly the wheel would turn. The impact water would not just roil in the blades and produce a wave, the gearing would have to be laid out in such a way (if the water wheel were to have this speed) that the cogs and staves would end up breaking off in an hour. Consequently, it is obvious that most of the theoretical calculations are rarely applicable in practice.

Part 5

In order to determine the correct gear ratio for reduction gearing according to a rule of thumb that is understandable to the practitioner, we have to derive the ratio from a mill that works well. Using the appropriate calculation, we can determine the correct gearing between the wheels for many different conditions. Let's say that we have 10 inches of fall and 100 cubic feet of water per second as the flow rate for a panster mill. Under these conditions, we can build a panster mill with the following proportions.

We make the water wheel 8 ells high and give the spur gear 65 cogs spaced 5 inches apart. The lantern gear gets 32 staves, the pit wheel gets 61 cogs spaced 4½ inches apart, and the stone nut gets 8 staves. With these proportions, the millstones have to have a diameter of 1 ell, 12 to 14 inches. Converting these proportions to revolutions, the millstones will turn 15½ times for one revolution of the water wheel, which is the right number of revolutions for millstones when a panster water wheel is 8 ells high.

1) Note

The height of the panster water wheel is determined by the same rule of thumb used for Staber mills, as stated in Chap. 2, Part 11. For panster water wheels, however, we have to use a different rule to determine the gear ratio, as I will explain in the next section.

Part 6

With the correct gear ratio, as described in the previous section, and with an 8-ell-high panster wheel, whose height is also based on the assumed fall and water quantity, we can deduce the following rules for many different fall heights and the water quantities required for panster machinery:
1) For each foot of height of a panster water wheel, we always give the millstone one revolution when the millstone has a minimum diameter of 1 ell, 12 to 14 inches. For example, if the panster wheel were 7 ells (or 14 feet) high, then the millstone would have 14 revolutions if it had the aforementioned diameter.
2) If the type of milling done in a region is such that the millstones can be bigger than the minimum diameter of 1 ell, 12 to 14 inches, as mentioned in Chap. 2, Part 14, then you can base the number of revolutions on the more usual diameter, as given in Chap. 2, Part 14. Namely, millstones with a larger diameter have fewer revolutions.
3) After determining the height of the water wheel based on the measured fall, you figure out the relative height of the rest of the gears according to the height of the water wheel. The relative heights are determined using the following rules, which are based on actual observations and proven experience. The diameter of the spur gear is always half the height of the water wheel. Then add 4 more cogs to the periphery of the spur gear, whose circumference is based on this diameter, to obtain the correct gear ratio.

Furthermore, the radius of the spur gear is the entire diameter of the lantern gear. Use the diameter to determine the circumference, and again add 4 staves to the periphery. This is how to determine the size of the lantern gear, and the spacing for the spur gear determines the number of staves that the lantern gear has to have. The pit wheel, however, gets 4 fewer cogs than the spur gear. Finally, the stone nut[45] gets as many staves as those on one quarter of the lantern gear. Using this method, the gear ratio can be determined for all different water wheel heights, as the following examples will show more clearly.

2) Note

With panster machinery, you always stay with 5 inch spacing on large gears and 4½ inches on small ones. Using smaller spacing for twice the load is useless and is only an imaginary game played by some practitioners.

Part 7

Let's clarify the proportionality method for gearing using an example. Let's say the height of a panster wheel based on its fall height were 7 ells, 12 inches. The radius of this water wheel would be 3 ells, 18 inches or 90 inches. From the radius, find the circumference with the following formulation:

7 : 22 = :90
       22

   180
   180

 1980

[Long division on p. 120 of German original]

282 and 6/7 inches is the circumference, but we use 283 inches to avoid the fraction.

Divide the above circumference by the 5 inch spacing to get the number of cogs for the spur gear minus[46] 4 (as described in Part 6), as follows:

[Long division on p. 120 of German original]
56   cogs
 4   in addition, thus
----
60   cogs

is the number for the spur gear, which we assume to be 61 cogs due to the alternation[47].

---

[45] Translator's note: The author uses the term *Getriebe* in this sentence, which is a general term for *gear wheel*. *Getriebe* is translated in the *Draft Dictionary of Molinology* as both *lantern gear* and *stone nut*. However, the author uses *Drehling* for a lantern gear, so by process of elimination, I am translating *Getriebe* as *stone nut*.

[46] Technical annotation: Apparent error in the original German text; the procedure calls for addition not subtraction. It is shown correctly in the example that follows.

2) The radius of the spur gear is 45 inches. Consequently, the circumference[48] is 141 inches (the fraction is not included, because it is unnecessary). Divided by 5 inches, the quotient is 28.

Formulation:

[Long division on p. 120 of German original]

28

4   in addition, thus

----

32   staves that are used for the lantern gear

Now the pit wheel gets 4 fewer cogs than the spur gear, namely, 57. The stone nut gets as many staves as those on one quarter of the lantern gear, namely, 8. This is how to proportion the gearing.

Part 8

If we now describe the gear ratio in terms of the number of revolutions, the quotient comes to 13 and almost ¾ revolutions of the millstones. This calculation would be as follows:

Formulation:

            61   cogs, spur gear
            57   cogs, pit wheel

          427
          305

        3477   first product.

32   staves on the lantern gear
8   staves on the stone nut

----

256 second product.

[Long division on p. 121 of German original]

$13\frac{149}{256}$ revolutions of the millstones

This number of revolutions of the millstones for one revolution of the water wheel is correctly proportioned for the height of the water wheel and for 1 ell, 12 to 14 inch diameter millstones, as mentioned previously.

---

[47] Translator's note: The term *Wechslung* can mean change, exchange, alternation, variation, displacement, etc. but I'm not sure what the author means in this sentence. Perhaps he is referring to adding one tooth to get the correct speed ratio.

[48] Technical annotation: The circumference refers to that of the lantern gear.

Part 9

The rule mentioned in Part 13 of Chapter 2 regarding the gear change also applies to panster machinery. This rule says that for the normal water level required to build the mill, one revolution of the millstone is subtracted in proportion to the height of the water wheel. When the water level is low or high, this revolution is compensated for with a stone nut that has one fewer stave. If you follow this rule, you will not fail to achieve the desired goal.

Part 10

The rules of thumb that you learned in Part 6 about determining the gear ratio also always apply in panster mills that use the largest diameter stones. For example, let's say that the type of milling in a region requires 1 ell, 20-inch diameter stones to be used. In this case, the number of revolutions of the millstones for the particular height of the water wheel is also based on the smallest diameter of the millstones and on the gear ratio. But then the calculation shown in Chap. 2, Part 14 is used to determine the correct number of revolutions for the larger-diameter stones. Because, by necessity, the larger-diameter millstones will have fewer revolutions as a result of the calculation, it is not necessary to reduce the number of cogs on the spur gear and the pit wheel based on the relative number of revolutions of the smaller millstones. Instead, give the lantern gears 4 or 8 staves and, similarly, the stone nuts 1 or 2 more staves than needed for the ratio using the small millstones to get the number of revolutions for the larger millstone. For example, in Part 7 we showed that the millstones turned almost 13 and ¾ times when the panster water wheel was 7 ells, 12 inches high. Let's say that you could use larger-diameter millstones with a water wheel with the same height, for example, having a diameter of 1 ell, 20 inches. We set up the calculation, just as in Chapter 2, Part 14, as follows.

Formulation:

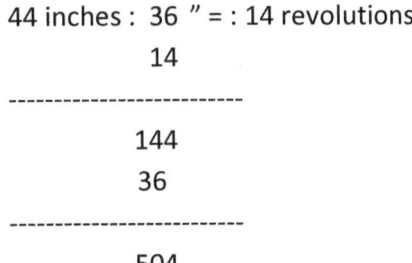

44 inches : 36 " = : 14 revolutions
            14

            144
             36

            504

[Long division on p. 124 of German original]
11 and 5/11 is the number of revolutions that the larger-diameter millstones (with a diameter of 1 ell, 20 inches) have to have. NB: the 13¾ revolutions is rounded to the next whole number.

If we use 34 staves instead of 32 for the lantern gear and 9 instead of 8 staves for the stone nuts, then we get the required number of revolutions, as shown by the following example.

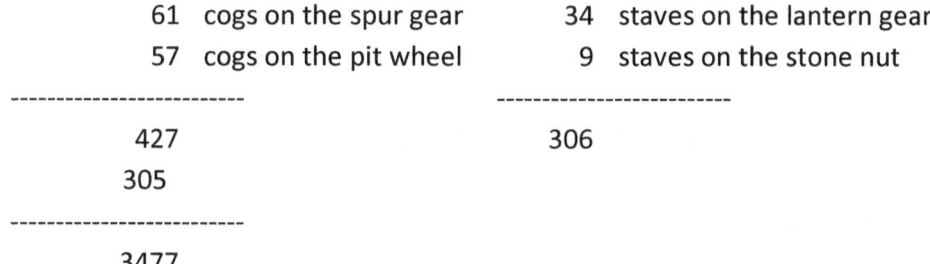

[Long division on p. 124 of German original]
11 and 37/102 or a little more than 1/3 is the number of revolutions of the millstone. The little bit that is still missing is not critical for the large machines.

Part 11

In the previous section, we described the rules according to which the proportions of the gears in panster mills are determined. Now we will turn our attention to the construction of the mills themselves, along with the layout of the individual components. Plates XII, XIII, and XIV are the drawings for a panster mill with 4 millstones that have the required water quantity for a fall height of 27 inches. If we lay the panster mill out on an apron, then the fall height will be partitioned as for a Staber mill, as described in Chap. 3, Part 2. In other words, divide the fall height into 9 equal parts based on the odd numbers 1, 3, and 5 and then proceed as follows: The upper water wheel A on Plate XIII receives 5 parts or 15 inches of fall height from the sill a to the cross-tree b. Then allocate 1 part or 3 inches of fall height from the cross-tree b of the upper water wheel to the apron sill c of the other water wheel B. That leaves 3 parts or 9 inches of fall height for the other water wheel. Using this procedure, the fall height is divided equally. We can lay the back portion of the race (as already mentioned for Staber mills) 5 to 6 inches deeper into the usual lower water level. In this way, we can give the second water wheel a few inches of fall more than it would get from partitioning the lively fall. This concludes the first step in building such a mill.

Part 12

The second task is to proportion the gearing according to the aforementioned rules. Based on the fall height mentioned previously, the gearing would be as follows:

The upper water wheel A on Plate XII is 7 ells high and has 28 blades on its periphery.

The spur gear B has 57 cogs spaced 5 inches apart. The pitch radius is 1 ell, 21 and 15/44 inches.

The lantern gears C C have 32 staves. The pitch radius is 1 ell, 1 and 5/11 inches.

The pit wheels D D have 54 cogs spaced 4½ inches apart. The pitch radius is 1 ell, 14 and 29/44 inches.

The stone nuts always have 7 staves.

With these proportions, the millstones revolve 13 and almost ¾ times for one revolution of the water wheel. The millstones generally have a diameter of 1 ell, 15 inches.

The other water wheel is 7 ells, 12 inches high and also has 28 blades on its periphery.

The spur gear *B* has 61 cogs spaced 5 inches apart. The pitch radius is 12 ells and 23/44.

The lantern gears *C C* also have 32 staves, the same as the upper ones.

The pit wheels *D D* have 57 cogs spaced 4½ inches apart. The pitch radius is 1 ell, 16 and 38/44 inches. With these proportions, the millstones revolve 15 and almost ½ times for one revolution of the water wheel.

These are the proportions of the gears, which determine how the rest of the mill is built.

Part 13

Once we know the height of the gear wheels, the third task is to determine the length and the position of the shafts. First, to determine the length of the water wheel shafts (Plate XII):

	Ells	Inches
For the distance between the crosshead *a* and the head stock *c*	---	4
For the thickness of the head stock *c*	---	12
For the width of the side wall *d*	---	8
For the width of the water wheel *A* including the clearance between the side walls	4	2
For the width of the second side wall *d*	---	8
For the thickness of the inner head stock *c*	---	12
For the distance between the inner head stock and the wall *e*	1	---
For the wall thickness	1	12
For the distance between the wall and the face of the lantern gear shafts *f f*	---	14
For the length of the neck of the lantern gear shafts	---	16
For the spacer up to the lantern gear arm *g*	---	12
For the width of the entire lantern gear *C* including the arms	1	12
For the space between the inner lantern gear arm and the lifting frame *h*	---	16
Sum of the length of the panster shafts	12	8

The length of the lantern gear shafts is based on the following components:

	Ells	Inches
For the length of the neck of the shaft *i* from the end of the neck to the lantern gear *c*	---	16
For the spacer up to the lantern gear arm *g*	---	12
For the entire width of the lantern gear *C* including the arms	1	10
For the space between the lantern gear arm and the lifting frame *h*	---	16
For the thickness of the lifting bay uprights and the lifting frame *h*	---	18
For the thickness of the sill beam *l* including the space to the lifting bay uprights	---	18

	Ells	Inches
From the sill to the pit wheel D	1	12
For the thickness of the pit wheel D with the arms	---	14
For the spacer up to the neck	---	10
For the length of the neck	---	18
Sum of the length of a lantern gear shaft	8	---

1) Note

When calculating the length of the lifting shafts, it is not necessary to take into account the horizontal components over which they lie. Instead, make the lifting shafts longer than the panster wheel shafts by the thickness of the lifting bay uprights and frame strips, which, in this case, is 1 ell, 12 inches on both sides. Consequently, the length of the lifting shafts is 13 ells, 20 inches.

Part 14

The fourth task is to determine the length and width of the building. First, the width is calculated as follows:

	Ells	Inches
For the thickness of the water wall e	1	12
From the wall to the face of the lantern gear shaft i	---	12
For the length of the lantern gear shaft ff	8	---
For the thickness of the pit wheel shaft bearing m	---	10
For the length of the bolting hutch n	3	---
For the space between the bolting hutch and the partition o	3	---
For the width of the mill room to the land wall p	3	12
For the thickness of the land wall p	1	6
Sum	21	4

This is the width of the building including the walls. In addition, the length of the building is calculated as follows: The weir sill is the reference point and is used as the standard for determining the horizontal position of all shafts as follows:

	Ells	Inches
From the inner side of the weir sill E on Plate XII to the periphery of the upper water wheel A	2	6
For half the height of the upper water wheel A	3	12
Sum	5	18

to the middle of the upper water wheel shaft. To make sure the gable wall p is the correct distance from the middle of the water wheel shaft, the dimensions of the following components are taken into account:

	Ells	Inches
For half the height of the upper spur gear B	1	22
For the pitch radius of the upper lantern gear C	1	1½
For half the height of the upper pit wheel D	1	19
For the space to the brayer beam r	---	8
For the thickness of the brayer beam	---	4
For the projection of the bridge tree s	---	6
For the space between the bridge tree s and the wall p	1	6
Sum	6	18½

from the middle of the upper water wheel shaft to the gable wall *p*.

Now we give the dimensions of the components from the middle of the upper water wheel shaft in the other direction as follows:

	Ells	Inches
For half the height of the upper spur gear B	1	22
For the pitch radius of the other lantern gear C	1	1½
For half the height of the other pit wheel D	1	19
For the space between the pit wheel and the last brayer beam for the other pair of millstones	---	8
For the thickness of the brayer beam r	---	4
For the projection of the last bridge tree for the other pair of millstones	---	6
From the bridge trees to the middle of the solepiece t	---	16
Sum from the middle of the upper water wheel shaft to the middle of the solepiece t	6	4½
Adding the aforementioned sum of	6	18½
Sum	12	23

is the space for the two top mills not including the gable wall *p*.

To determine the position of the second panster or water wheel shaft as well as the correct positions of the remaining required components, we continue with the dimensions from the last point as follows:

	Ells	Inches
From the middle of the solepiece t to the first bridge tree of the third pair of millstones	---	16
For the projection of the bridge trees	---	8
Sum	1	---

	Ells	Inches
Sum carried forward	1	---
For the thickness of the brayer beam r	---	4
For the space to the 3rd pit wheel D	---	8
For half the height of the 3rd pit wheel	1	21½
Sum to the middle of the 3rd pit wheel shaft	3	9½
For the pitch radius of the 3rd lantern gear	1	1½
For the entire height of the 2nd spur gear B	4	2
For the pitch radius of the 4th lantern gear	1	1½
For half the height of the 4th pit wheel	1	21½
From the 4th pit wheel to the last brayer beam	---	8
For the thickness of the brayer beam	---	4
For the projection of the last bridge tree	---	8
From the last bridge tree to the wall	1	6
For the thickness of both walls	2	12
Sum	16	2
Adding the space for the two upper mills	12	23
Sum	29	1

is the length of the entire building including the walls.

Part 15

Now that we've determined the dimensions of the components in the horizontal plane, let's turn our attention to their layout in the vertical plane, from which the height of the hursting and the building is calculated. This is the fifth group of components that has to be calculated. The upper cross-tree is used as the reference point and the components are shown on Plate XIV.

	Ells	Inches
For the radius of the upper water wheel A	3	12
For the height of the lantern gear shaft f above the middle of the water wheel shaft. NB: when the water wheel is hanging on the ground	---	12
For half the height of the pit wheel D	1	19
For the space between the pit wheel D and the beams of the hursting u	1	---
For the thickness of the beams u	---	4
For the height of the millstones v	1	12
For the clearance between the hopper x and the millstone v	---	5
For the height of the hopper x	2	---
For the space between the hopper and the beams y	2	12
Sum	13	4

from the cross-tree of the upper water wheel A to just below the beams of the building y.

Measuring 18 inches down from the middle of the pit wheel gives the position of the floor and the sill beam *l l*. Now we can easily calculate the hursting dimensions starting from the floor:

	Ells	Inches
For the height of the pit wheel shaft f above the floor l	---	18
For half the height of the pit wheel D	1	19
For the space to just under the beams u	1	---
Sum	3	13

is the height of the hursting excluding the thickness of the beams. These are the most important components of such a mill. The rest of the layout in a panster mill can be seen clearly in the drawings and their dimensions can be determined using the scale bar.

Part 16 Lifting Gear

Lifting gear, as you know, is used to adjust water wheels according to changing water levels. Determining the proportions of the lifting gears relies on mechanics. The derivation of this formula does not belong here, but I will try to explain in understandable terms how to apply mechanical principles to mill and machine construction in a special book for practitioners. For now, however, I will just describe the gear ratio of a lifting gear, because this also has to be included in the design of a panster mill. The lifting spur gear (see Plate XIV) has 80 cogs spaced 5 inches apart. The pitch radius is 2 ells, 15 and 7/11 inches. The collar 2 has 6 staves. The Y-wheel 3 has 36 staves and is 4 ells, 12 inches high. The ratio of power to load is almost 1 to 80 for this gear ratio in equilibrium. In other words, if you hang one zentner on the outermost stave of a Y-wheel so that it is perpendicular to the horizontal plane, then this weight would be balanced with 80 zentner on the lifting chain.

I have already explained in detail in Volume 1 of my book on the art of building mills how to construct these wheels to the greatest advantage. Therefore, it is not necessary to say anything about it here.

Part 17

I feel that the following topics about lifting gear deserve to be mentioned:
1) For lifting gear to be positioned correctly, the center lines of the water wheel shafts must be parallel to the center lines of the lantern gear shafts. Otherwise, not only will it be hard to pull up the lifting gear, the spur gear will be noisy.
2) The uprights *a a* on Plate XIII must stand perfectly vertically on the inside and the outside and they must be protected against vibration if the spur gear is to run quietly. To achieve this purpose, the uprights *a a* on Plate XIII must be inserted in strong sills on the outside and must rest on strong stone slabs on the inside where the spur gear is located. This can be seen clearly in the side view on Plate XIV. Horizontal supporting beams are cogged onto one end of the gable beams *d d*. The other end is anchored securely in the water wall. To prevent the lifting frame from swaying back and forth when the lifting machinery is running, the lower cross beams *e e* protrude inward 3 inches from the uprights. In this way, the lifting frame can be raised and lowered as in a groove. The rest of the lifting gear components can be seen clearly in the drawing.

Part 18

To give a massive mill building the appropriate stability, a frame must be laid below the water wall so that it does not sink or burst. This frame must not only be laid under the entire stretch of the water wall, but also for a stretch under both gable walls into the bank. The construction of such a frame is described below.

For the length of the wall, sills *a a* (Plate XIV, Fig. 2) are laid on strong piles at a distance from each other that allows the frame to protrude beyond the wall one ell on each side. Furthermore, these sills are connected to solepieces *b b* and so on, as shown in the drawing. The resulting compartments *c c* are densely filled with alder, or with alder piles, which are driven into the ground as far as possible.

After the frame is finished thus far, two layers of work pieces are laid on the frame, as shown on the ground for the water wall H on Plate XIV. These work pieces are laid out so that the first layer of stone is even with the top of the frame sills *a a* in Fig. 2. The second layer is 3 to 4 inches farther in from the edge, but still wide enough for the foundation of the actual wall to rest on it completely. The wall thickness is then reduced 18 or 20 inches and built with this width (here 1 ell, 12 inches) the rest of the way up to the roof.

Part 19

If, however, the ground on which the water wall is to be built is not too loose, then the ground can be stabilized without building a frame, with less expense, as follows: Cover the lower ground with rubble, tamp it down with a manual pile driver, and lay two or three layers of bricks on the bare sand so that the bricks form a level surface. Build the wall on these layers of bricks. If a brick is crushed into this layer of sand, the wall, which is held together with lime, will not crack for the following reason. The broken or displaced brick on the layer of sand only supports a portion of the wall. However, it is always better to support such heavy water walls for mills with a frame, not sparing the expense.

Part 20

There is an important consideration when building panster mills, which deserves to be explained here. This consideration pertains to selecting the appropriate measures to prevent such a mill from becoming unusable during flood conditions. How many well-built mills become unusable as soon as there is a moderate increase in water level, so that obstruction of the wheel causes not only the mill owner, but also a portion of humanity to suffer loss and damage? To prevent rising water levels from obstructing the function of not just panster mills, but every other mill, take care to observe and carry out, as much as feasible, the following recommendations when building a mill.

1) Always try to build the outlet channel[49] below undershot mills so that, whenever possible, it continues in a straight line for 1500 to 2000 paces from the tail race before there is a bend. It is worth it to come to an agreement with the property owners through whose fields and meadows you have to cut to get to a straight section of the river, if the local conditions don't make this completely impossible. In this way the water is discharged from the tail race into the river in a very lively manner, so that the water wheels don't choke so easily.
2) If it is necessary to build an overflow weir for a mill, then I warn every master millwright about the main thing he must be aware of when building a weir. Do not let the weir channel impact the mill channel too closely downstream of the mill. If possible, let the weir channel join the mill channel 2 to 3000 paces downstream of the mill, to protect the mill during flood conditions.
3) Try to keep the mill channel downstream of the mill free of impurities caused by uneven areas in the river bed. At normal water levels, these impurities do not cause the water to back up and prevent the water wheel and associated machinery from turning. However, this happens right away with the slightest increase during flood conditions.

3) Note

Based on personal experience, these are the best ways to keep a mill going as long as possible during flood conditions. One mill, 3 miles from Leipzig, was fortunate enough to keep on working even when all the other mills in this region became unusable due to flooding. Even during the flood of 1784, when almost all the mills around Leipzig and Halle were just standing around, this mill was working at full capacity! Why, do you ask, was this particular mill not hindered and rendered unusable by the flood? Here is the reason: The channel downstream of the mill, which discharged the water, was 2000 paces long in a straight line from the mill race, before it had a bend. Furthermore, the water from the spillway was discharged into another straight outlet channel, whose discharge entered the weir channel. After 2500 paces, the weir channel merged with the river or mill channel. Because these channels ran in a straight line downstream of the mill races, the water had a high discharge. If this discharge was fast in the mill channel, then it necessarily had to stay fast in the wheel race so that the mills could work at full

---

[49] Translator's note: The author differentiates between *Graben* and *Gerinne*, which both mean *channel*. My sense is that a *Graben* is an unfortified channel (a ditch), while the *Gerinne* is man-made (a mill race).

capacity even when the water had risen vertically over one ell. The owner of this mill once allowed the channel to silt up, because he thought it wasn't necessary to remove the silt. The old guys tricked him out of ignorance into thinking that he could make better use of the area when the channel was full. This plan seemed to be economically sound, but what were the consequences? Every time there was a flood, more and more water backed up in the mill race. Within two years, when there was just a moderate increase in the water level, the second wheel for the lower millstones could no longer function as it should. Consequently, the outlet channel had to be cleared, which then improved the situation so that when the water in the river rose vertically by one ell, the level still did not rise under the cross in the third water wheel, which drove the oil mill.

4) Note

For panster mills, the hursting must be such that there is at least one ell of clearance between the pit wheel and the post. This is to allow the wheel to be lifted even during flood conditions, as Melzer and Beier also write in their mill books. When the mill channels are not straight, however, as mentioned in Part 20, this measure doesn't help much or at all. It doesn't matter how high the wheel is lifted when the water velocity in the race is zero; the mill will not be able to function.

Part 21

If the local circumstances absolutely do not make it possible to run the mill channel below the mill in a straight line for the length specified in Part 20, then you can use the following alternatives to still keep your mill running during flood conditions.

1) Use lantern gears with double staves, whose construction is described in Volume 1 of my book on the art of building mills, Chap. 6, Part 15. During flood conditions, knock out the outer staves so that the spur gear meshes with the inner staves.
2) Nail boards onto the arms of the water wheel, so that they protrude 2 ells from the shrouds. These boards increase the surface area for the slowly flowing water to impact. This is another important way to help a mill keep running during flood conditions, as I know from personal experience.
3) All mills that are at risk for becoming nonfunctional during flood conditions must have very good, waterproof walls so that the gearing does not easily become wet. If these rules would be followed when building a mill, for sure more mills could remain functional for a longer time during floods.

Part 22

Low water levels for a prolonged period are also bad news for our mills. Therefore we are also bound to think of measures that allow our mills to maintain good functionality for as long as possible. For panster mills in particular, the following rules are recommended.

1) When you have discovered through inquiry that the low water periods are significantly longer than the flood periods for a given river, then never build the wheel race for a panster water wheel wider than 4 ells. A wide race that does not contain the right amount of water for the gearing will not function well when the water level is low. The reason is that a small amount of

water flows at a low level in a wide channel and the water cannot have the same force on the blades of the water wheel as in a narrow channel, where the water level is higher.

2) When the water level is low, nail guide boards into the wheel race, as shown on Plate VI for a Staber mill. These boards force the water into a smaller space, thereby increasing the impact on the blades.

3) It is important to maintain the gearing in good condition so that you don't waste too much power on friction. To avoid friction on cogs and staves, see Volume 1 of my book on the art of building mills for instructions for building good spur gears. Here I just want to mention that the gudgeon should turn on little metal rods, whose manufacture is already familiar to most millers. The stone spindle also creates a lot of friction when the steel does not have the appropriate hardness. To get the appropriate hardness, let the stone spindle and the bearing rods harden in Ulaunwasser[50]. When running, they get a very smooth surface and they also do not heat up.

4) Milling the grain and tentering the millstones are very important, not just for obtaining good flour but also for promoting mills.

Because it is not my intention in this book to give instructions on how to mill grain or tenter millstones, however, I will write a special publication for the mill audience on these topics in the future. I will describe how to mill every kind of grain so that you can obtain not just the finest, but also the most prevalent flour, and similarly, how to make the fine English flour, etc. I would hope that you don't believe that grinding is a simple task, yet how little attention is paid in many mills to this important matter for humanity. Many a poor person has toiled to produce a little bit of grain, which he entrusts to be ground not by an actual mill hand, who has properly learned his profession, but often by apprentices and donkey drivers.

---

[50] Technical annotation: We have found no direct translation for the word "Ulaunwasser." However, the context indicates a recommendation to harden the spindle and bearing rods in some type of water. A close current German translation is shown as "aluan", which is a chemical material, alum. Alum in water acts acidic, and hence could have been used as a pickling agent to clean the surfaces of the spindle and bearing rods to help reduce friction during running operation of the mill equipment.

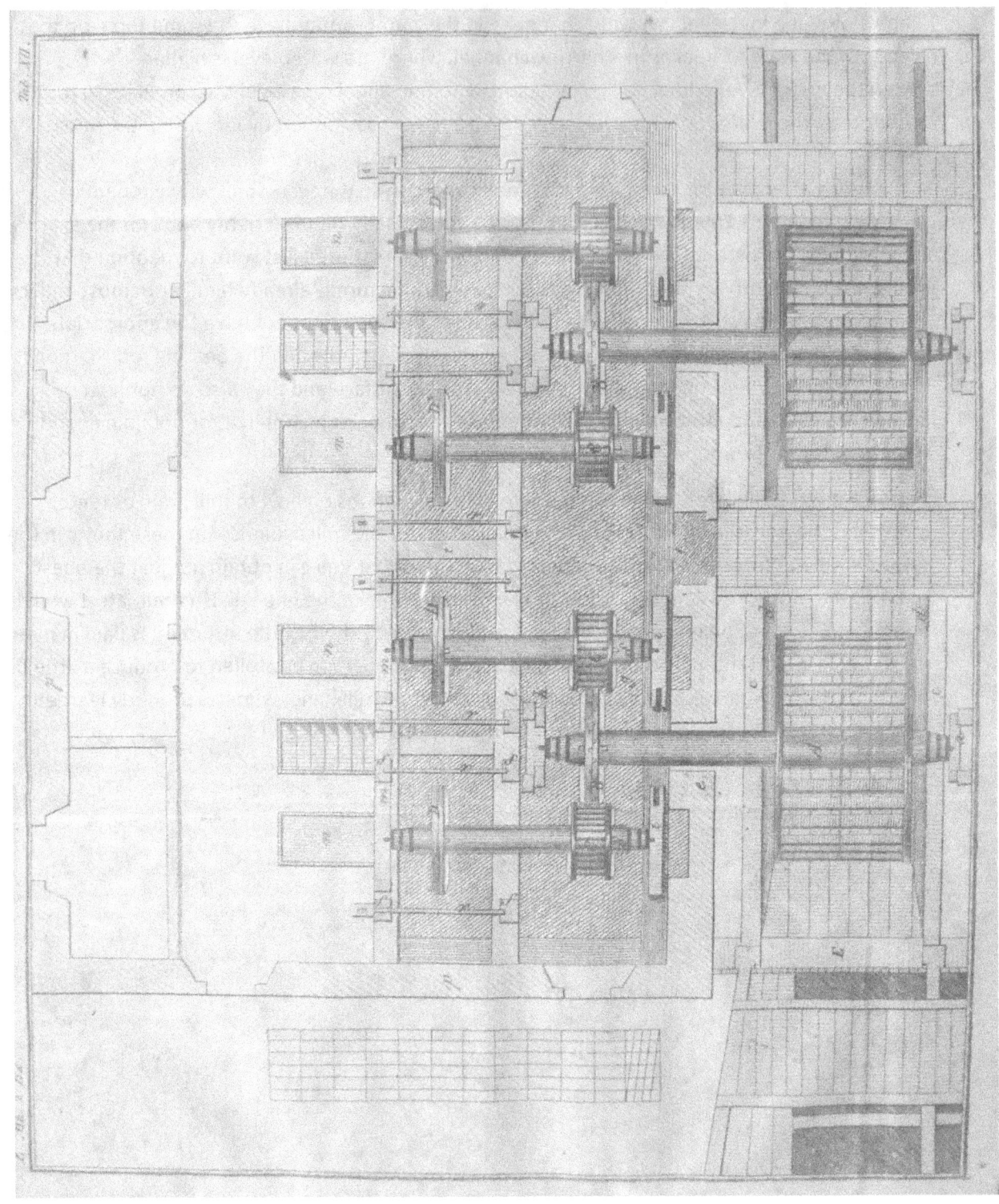

Plate XII

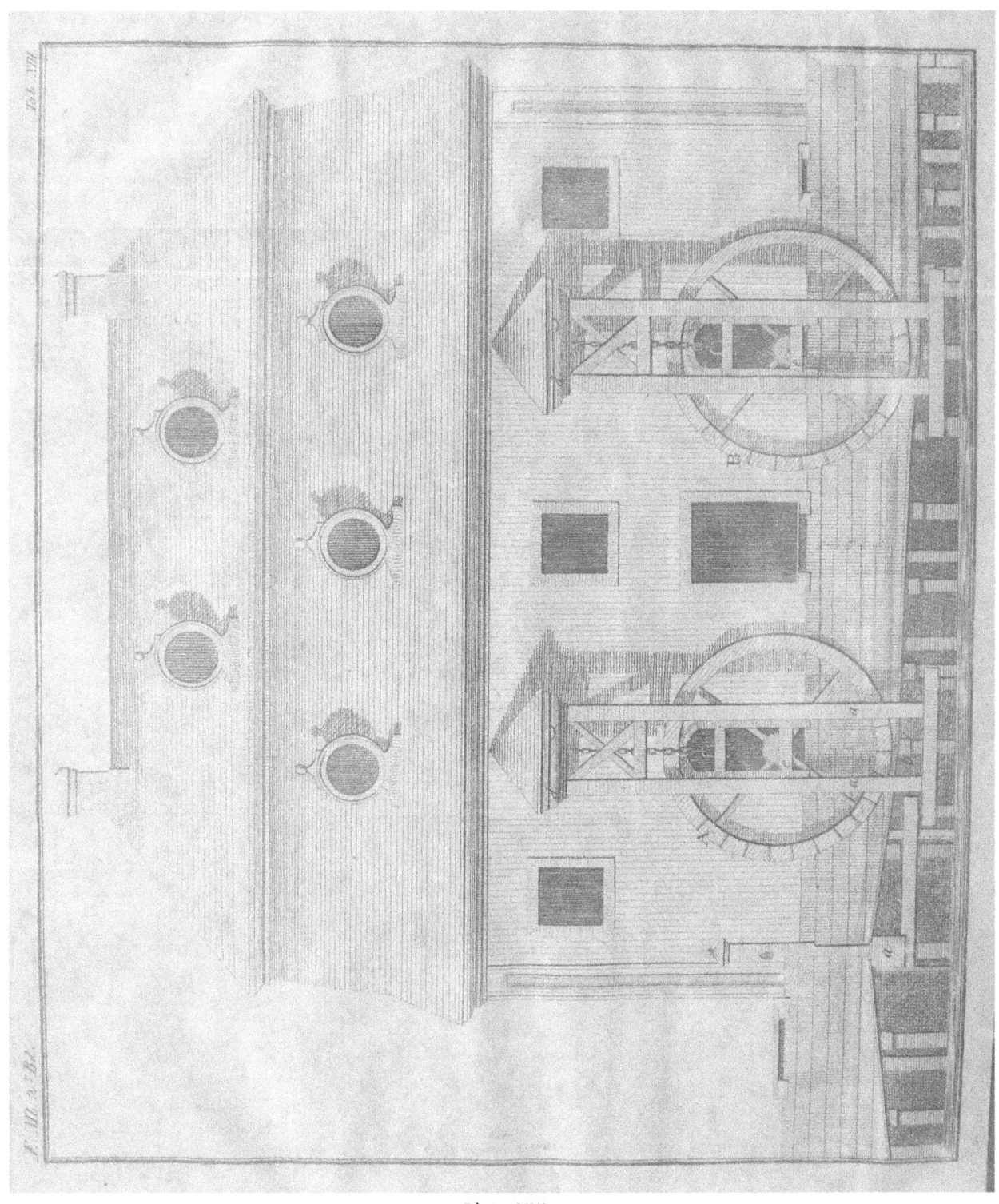

Plate XIII

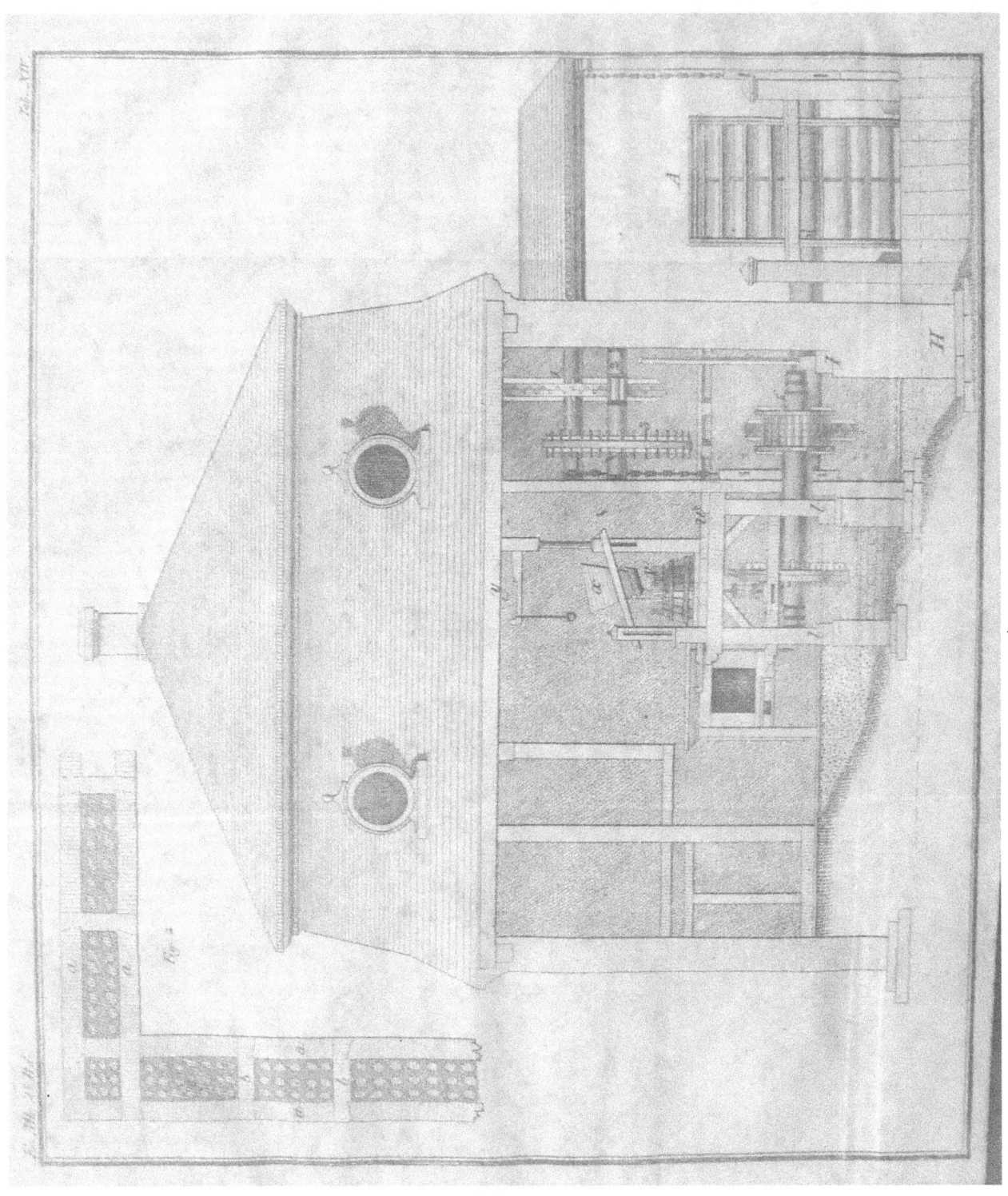

Plate XIV

Chapter 7

Building a Strauber Mill with Two Millstones

Part 1

Strauber[51] water wheels are appropriate for fall heights that are somewhere between those needed for Staber wheels and overshot water wheels, as mentioned previously in Chap. 2, Part 8. Based on the rules of thumb given in Chap. 2, if a river does not supply enough water to build a mill with a Staber water wheel, then the next option is to build a Straub water wheel.

Part 2

The most important rules of thumb for building a Strauber mill are the proportions of the gears and the proper layout of the aprons for the water wheels. I will show how to calculate the proportions of the gears for Strauber mills in the next section. Regarding the aprons, I have already shown clearly how to lay them out in Volume 1 of my book on the art of building mills. Therefore, it is not necessary to repeat my instructions here.

Part 3

When building a Straub mill that drives one pair of millstones, the revolutions of the millstone and the proportion of the gears are determined using the same rules that were given in Chap. 2, Section 13 and so on for building Staber mills with one millstone. For Straub machinery, however, there is an exception. Always give a Straub wheel that has the same height as a Staber wheel two more millstone revolutions than the gear ratio for Staber machinery requires. Then you get the correct ratio for Strauber gearing. For example, a millstone would turn 12 times for each revolution of an 8-ell high Staber wheel, as shown in Chap. 2, Part 11. For a Strauber wheel of the same height, the millstone would turn 14 times. From this ratio, it is possible to use the previously described rules for Staber wheels to calculate the gear ratio for Straub wheels.

Part 4

If there is a good reason to build Straub machinery with reduction gears, the gear ratio is determined as described in the previous chapter for panster mills, with the one exception. For a Straub water wheel, always add two revolutions of the millstone to the number resulting from the gear ratio for the panster water wheel, assuming that the millstones have a diameter of 1 ell, 10 to 14 inches. If the fall height for a Strauber mill is 1 ell, 12 to 14 inches, and the river delivers 13 to 15 cubic feet of water per second, then the millstones can be 1 ell, 13 to 14 inches in diameter. With this fall height and water quantity, the millstones can be given two revolutions more than the number resulting from the gear ratio for a panster water wheel that has the same height as a Straub water wheel. For example, let's say a Straub

---

[51] Translator's note: The author uses the terms *Strauber* and *Straub* interchangeably. Based on the drawings, a Straub water wheel has the blades mortised into the shrouds, but they project beyond the shrouds (see Plates XV and XVI). In contrast, the blades of a Staber wheel are flush with the face of the shrouds (see, for example, Plates IV, VI, and VIII).

water wheel is 8 ells high and you want to build a mill with reduction gearing. Based on the gear ratio, a panster water wheel having the same height would cause the millstones to revolve 15 times. On the other hand, the millstones would revolve 17 times with a Straub water wheel. If you have to build a Strauber mill where there is little water, however, you would do well to make the millstones only 1 ell, 12 inches in diameter and to give the millstones 3 revolutions more than those of a panster mill with the same gear ratio. Under these conditions, you can expect the mill to have the best possible effect.

Part 5

When laying out reduction gearing for Straub machinery, pay attention to the following rules if you want the mill to function well.

1) As mentioned for panster mills, always make the spur gear half the height of the water wheel. However, instead of adding 4 more cogs to the periphery, as you would do for panster wheels, add 8 cogs for Straub wheels. Furthermore, with panster wheels, the pit wheel had 4 fewer cogs than the spur gear. Similarly, with Straub wheels, give the pit wheel 8 fewer cogs than the spur gear. This layout results in the correct proportion of the spur gears and pit wheels for Straub machinery.
2) The number of staves on the lantern gear and the stone nut[52] is determined by the same rules that were described in the previous chapter for panster water wheels.
3) It is well known that the spacing on the gears that drive one millstone must be smaller than that which drives two. Therefore, for Staber, Strauber, or overshot water wheels using reduction gearing and driving one millstone, the cogs on the spur gear may not be more than 4 inches apart or, with a small water wheel, less than 3½ inches apart.

Part 6

To explain these rules clearly to my readers, let's apply them in an example. Let's say that we want to build a Strauber mill where there could be 1 ell, 12 to 14 inches of fall, but the stream does not supply the amount of water required by a Straub water wheel (namely, 13 to 14 cubic feet of water per second). We still want to build the little mill as advantageously as possible given the circumstances, so we would lay it out as follows.

Because the given stream does not deliver quite enough water for this kind of water wheel, the mill has to have reduction gearing to give the water wheel a slightly slower rotational velocity. In this way, water accumulates in front of the blades and more force is gained. Now let's say you found a fall height of 1 ell, 12 inches, which would require a water wheel 8 ells high. The gearing would be laid out as follows, based on the rules given in the previous section. Half the height of the water wheel is 4 ells or 96 inches, giving us the height of the spur gear. The circumference based on this height is 302 inches; divided by 4-inch spacing gives 75 cogs for a gear that has 4-inch spacing and this circumference. However, 8 cogs have to be added, which means that the spur gear gets 83 cogs. Furthermore, taking half the height (or the radius) of the spur gear as the height of the lantern gear, which is 2 ells or 48 inches, the resulting

---

[52] Translator's note: As mentioned in the previous chapter, *Getriebe* is a general term for *gear wheel*, which I am translating as *stone nut* based on context.

circumference would be 151 inches. Dividing this circumference by 4-inch spacing gives 37¾, but we assume 38 as the number of staves needed for the lantern gear. To determine the number of cogs on the pit wheel, subtract 8 from 83, which gives 75 as the number of cogs on the pit wheel. With this gearing, the spacing on the pit wheel is always a quarter inch less than that on the spur gear, which makes it 3¾ inches.

The number of staves on the stone nut is 9, which is always a quarter of the last even number of staves on the lantern gear. This is the manner in which the gear ratio is determined. If you calculate the number of revolutions completed by the millstone based on this gear ratio, the quotient is 18 and almost 1/7 times as the correct number of revolutions for the specified height of the water wheel.

Part 7

Determining the height of Straub water wheels has its limitations, as was also previously noted for Staber and panster wheels. The minimum fall height allowed for a Straub water wheel is 1 ell, 12 inches with 12 to 14 cubic feet of water per second. For this fall, an 8-ell high water wheel is required if the mill is to function properly. If the fall height increases 3 inches, then the water wheel can be 8 ells, 12 inches high. In other words, for each additional 3 inches of fall, the height of the water wheel always increases by 1 foot up to 1 ell, 18 to 20 inches of fall, where this type of water wheel reaches its limit. By using this empirical method, it is possible to determine the height of a Straub water wheel for every fall height.

1) Note

If a river supplies 12 to 14 cubic feet of water per second and the fall height is more than 2 ells, then it's better to build a flutter wheel. With that kind of fall, the flutter wheel is expected to be more effective than a Straub wheel.

How to build a flutter wheel most advantageously will be described elsewhere.

Part 8

Regarding the construction of a Strauber mill itself, this type of mill with two millstones, each driven by one water wheel, can be seen on Plate XV, which shows the plan view and on Plate XVI, which shows the side view from the water side. The layout of the mill is as follows:

The fall height here is assumed to be 1 ell, 12 inches.

The water wheels *A A* on Plate XV are 8 ells, 12 inches high to the ends of the blades. There are 48 blades on the periphery.

The pit wheels *B B* have 83 cogs, spaced 3¾ inches apart. The lantern gear gets 6 staves and the millstones are 1 ell, 12 inches in diameter. Converting this gear ratio to revolutions, for each revolution of the water wheel, the millstone turns 13 and 5/6 times, which is the correct number for a water wheel with the aforementioned height. Construction of the apron, shown on Plate XVI, follows the rules that I explained in Volume 1 of my book on the art of building mills, Chap. 4, Part 11.

Part 9

The other components of a Strauber mill are calculated relative to the position of the adjacent components in the horizontal and vertical planes, just as they were for simple Staber mills in Chap. 2 and 3. Therefore it is not necessary to explain their layout here; rather I will just name the components, in order, which are shown in the drawings. Plate XV shows the plan view with the components in the horizontal plane drawn to scale. 1) The main components are:

A A	The water wheels
B B	The pit wheels
C C	The bolting hutches
D	The steps
E	The mill room (or waiting room)
F F	The walls
G	The apron
H H	The two channels

2) The rest of the components are noted below:

a a	The side walls of the channel
b b	The bearings for the water wheel
c	The weir sill and the sluice frame
d d	The sill beams
e e etc.	The bridge trees of the hursting
f f	The brayer beams
g g	The studs
h h	The bearings for the pit wheel
i i	The solepieces for supporting the sill beams
k	The partition in front of the millstones

The rest of the mill layout can be seen more clearly in the drawing, where the position and the dimensions of all the components can be determined using the scale bar. Plate XVI shows the side view of the entire sluice, with the components in the vertical plane.

The layout of the water wheels *A A*, the weir sill *C*, and the framework with the piles and sills can be seen much more clearly in the drawing than I can describe in words.

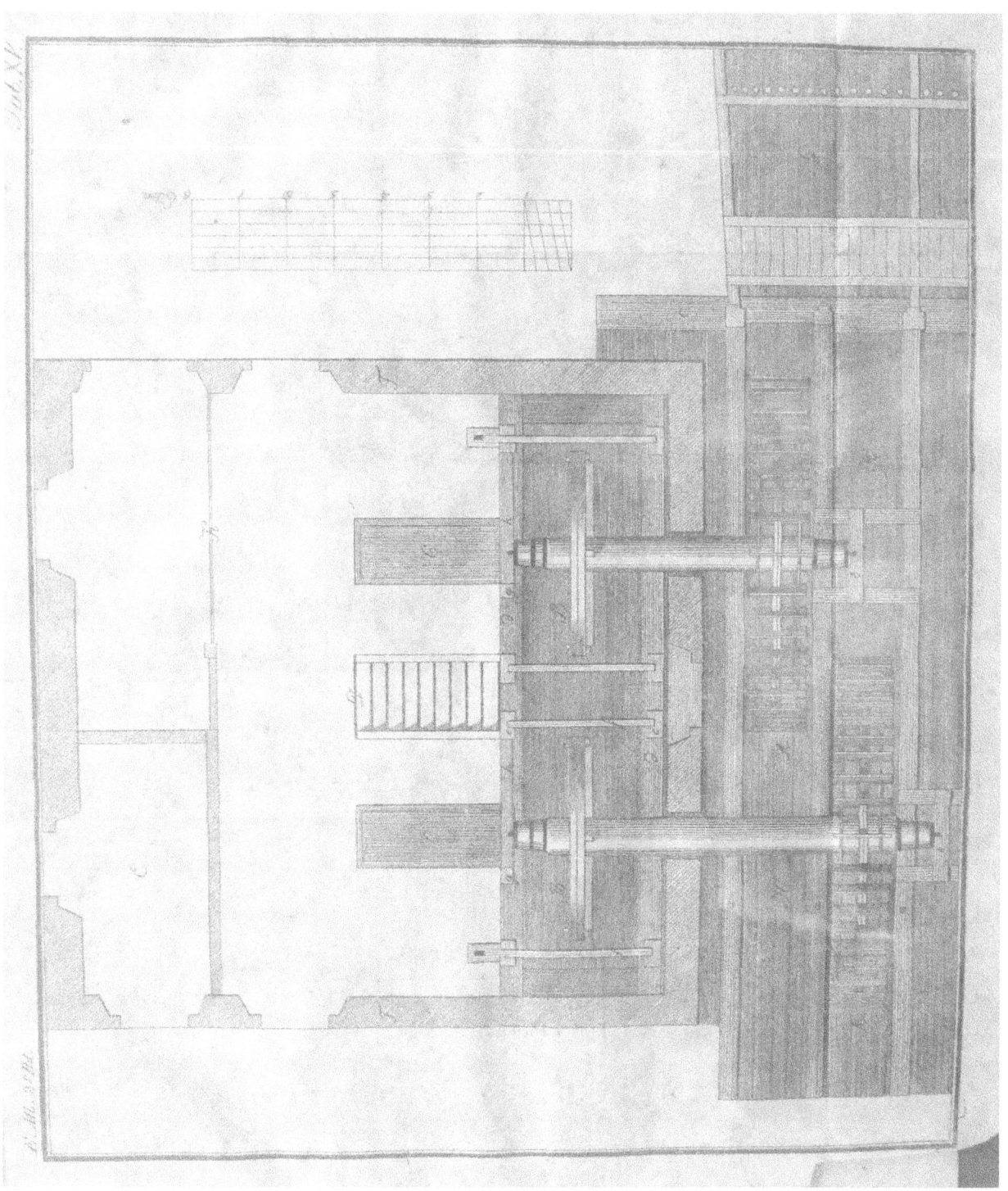

Plate XV

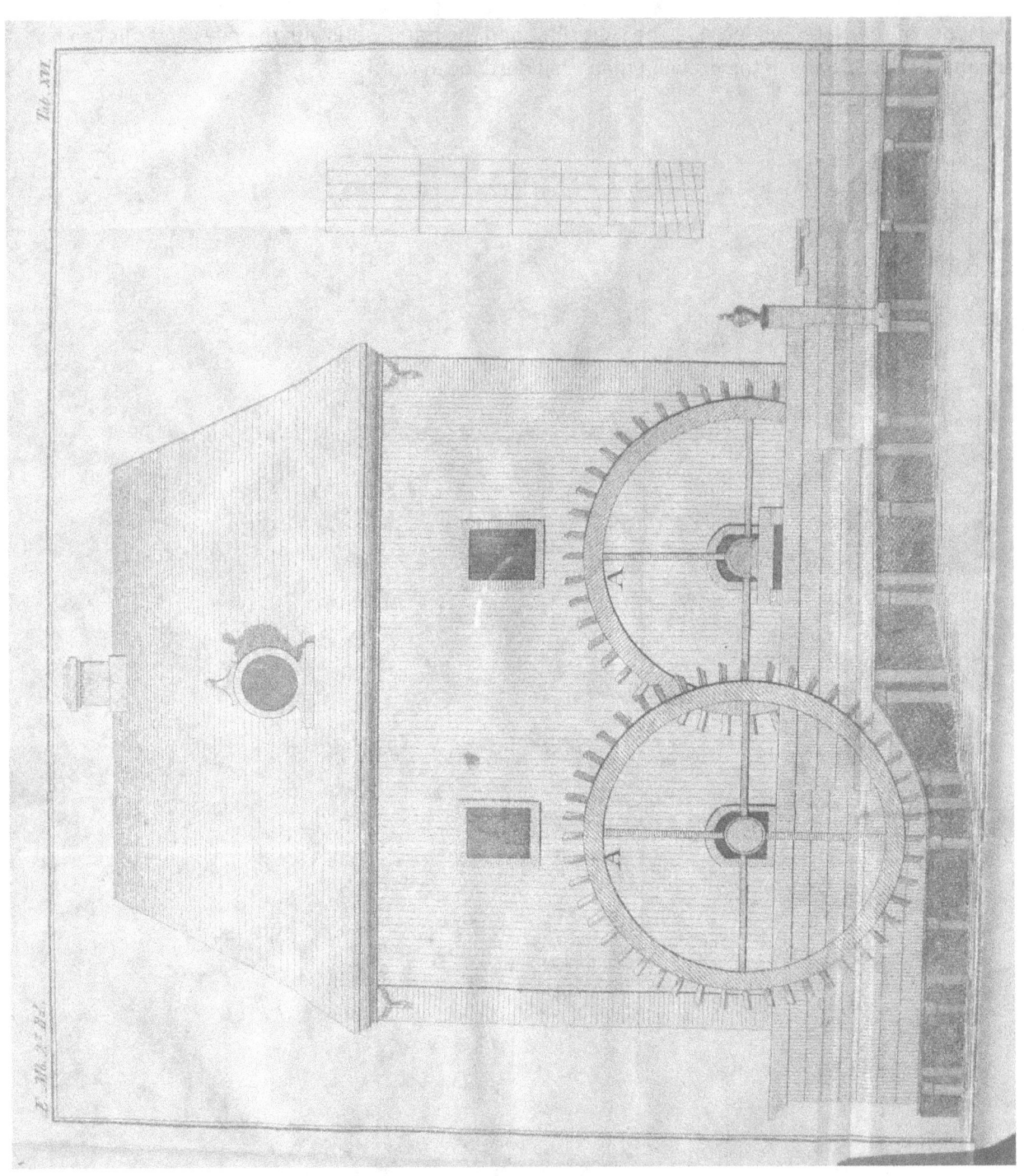

Plate XVI

Chapter 8

Proper Handling and Evaluation of Wood for Mill Construction

Description of an English Tannin Removal Machine

Part 1

Anyone who builds mills and machines and who works with these structures must be extremely knowledgeable about wood. He must know which type of wood to use for which purpose when building a mill.

The strength and durability of mills and machines depend a lot on the quality and the correct layout of the materials from which these structures are built.

Part 2

The wood used for mill gears that run dry, such as spur gears, pit wheels, lantern gears, etc. must be strong and well seasoned. Trees that grow on dry land are less susceptible to woodworm than those that grow in damp and swampy ground. As is generally known, young wood is softer than old wood, but when standing wood becomes too old, its strength decreases.

The heartwood of a tree is always stronger than the bark. For this reason, always choose good heartwood whenever possible for mill and machine components that have to withstand a lot of force. For the wheels[53], oak is the best among the deciduous trees, but make sure you choose Holm oak[54] and not English oak[55]. The English oaks have larger leaves and smaller acorns on longer stems; their bark is rougher than that of Holm oaks. In contrast, the wood from a Holm oak is oilier, finer, and also heavier than that of the English oaks. Oak is preferred in mills for the gear wheels, which run dry, for hursting, for shafts, for bay timbers and ice breakers, for foundation piles, and for many other components that are known to almost every master millwright. This kind of wood must always be in the same state of wetness or dryness; otherwise it warps more than other kinds of wood and also decays more easily. If it continually stands in water or where it is dry, then its natural hardness will further increase. If tannin is removed from the wood (this will be described in detail below), however, then the wood will not be subjected to warping.

Part 3

Red and white beech is a very hard and solid kind of wood when dry. Staves and cogs (pegs) for gear wheels and screws and other parts used in mills and machines are made from this wood. Pine is preferred for water wheels and piles that stand in wet ground. Alder is also a good wood for pilings. On the other hand, wood from elms is very useful for parts that always stand under water. Elm also

---

[53] Translator's note: *Räder* could mean gear wheels or water wheels.
[54] Translator's note: A Google search indicates that the Latin name for *Steineiche* is *Quercus ilex*.
[55] Translator's note: A Google search indicates that the Latin name for *Kohleiche* or *Raseneiche* is *Quercus robur*.

tolerates alternating wet and dry conditions, which is why I recommend it for the sheet piles of sluices and for the blades, particularly of overshot water wheels.

Part 4

Trees that grow on dry ground provide the best kind of timber, as I mentioned previously in Part 2. Here is the reason: Trees that grow on dry, sandy soil grow slower than those on damp ground.

Trees grown on dry and high ground are subjected more to storms and the weather, which generally cause them to become harder and better. The best time to fell timber is the beginning of spring, because then the wood, in which the sap is flowing, can dry the whole summer. On the other hand, sap in trees that are felled in the winter, stops flowing. This wood does not quite lose all of its sap, which would begin flowing again in the spring. Of course, the timber felled in the spring has to season for a while in order to dry out properly. Winter wood, on the other hand, never does dry out completely because it contains too much sap.

Part 5

Here are the most important rules that must be carefully observed when felling timber:

1) About 14 days before felling a tree, make a cut around the bottom of the trunk to draw out the sap.
2) About 8 days after felling the tree and removing the branches and the tree top, completely remove the bark. By removing the bark, the trunk will dry out easier and better. Furthermore, woodworms often live in the sapwood and then eat their way into the heartwood, causing the wood to be full of worms before it is used.
3) After the timber has been felled and the branches have been removed, you have to know how to test if the trunk is good throughout, or if there is decay inside, which cannot be seen with the naked eye. To determine if there is decay, use this procedure: Lay the trunk on a pair of supports and sharply hit the bottom of the trunk with a hammer. At the other end, listen for a clear, bright sound. If the tone is clear and bright, that is a sign that the trunk is good throughout and does not have any decay. On the other hand, if you hear a muffled sound, then the trunk is bad on the inside, and it doesn't matter where.

Part 6

When constructing mills and machines, it is undesirable to use unseasoned wood, because this wood shrinks and the fibers contract along their length. When the wood shrinks, it is liable to split. Shrinking in the direction of the fibers may take place, but shrinking in the direction of the heartwood (or perpendicular to the length of the fibers) is far worse. When the latter takes place, the lumber gets cracks all the way into the pith. To prevent this from happening, the trees (especially those that will be cut into boards) must be cut right away according to requirements before they are seasoned.

Part 7

To prevent timber for mill and machine construction from shrinking and cracking, take the following measures:

1) Do not dry the wood in direct sunlight, rather in a shady location protected from the wind so that drying does not take place too quickly.
2) The wood to be seasoned may not lie directly on the ground; rather, as is widely known, put supports under it so that the wood doesn't draw moisture.
3) It is even better if the ground on which the wood lies is covered with floorboards or pavers, because these cause the moisture coming out of the ground to be held back even more.
4) If the conditions allow, the wood will dry even better when it is covered with dry sand. Finally,
5) the best method involves removing tannin from the wood. This process is done 6-8 weeks before it is used, either by laying the wood into a flowing stream or by putting it into a tannin-removal machine, which will be described in the following section.

Part 8

When the wood is dried in sand, the outer pores remain open, because the sand, which absorbs the moisture in the wood, keeps the surface moist. Because of this, the warm moisture on the inside can also evaporate and thus the wood can dry out completely. One would do well to hang up small parts that are required in mills and machines, which must still be hard, such as, for example, cogs (pegs) and staves, in a chimney for a few weeks where the smoke always goes through. This treatment causes the parts not just to become hard, but also to dry out completely without cracking.

Part 9

Wood treated in this manner is completely dried out, which can be recognized by the fact that small cracks arise in the pith. These cracks become evident because the pith dries first, followed by the rest of the wood towards the surface in the course of steady, gradual drying. These small cracks do not damage the wood, however, because they do not run in a straight line. Rather they leave intervals of uncracked areas of wood between them. The presence of these small cracks in the pith is the surest indication that the pith is dried out. To determine whether the outer areas are also dried out, check if the sand adjacent to the wood is still damp to the touch. If it is, then the wood has to stay in the sand a little longer until the sand no longer feels damp. If, however, the wood must retain some moisture for a specific reason, then the wood can be removed from the sand earlier, not waiting for the appropriate drying time to elapse.

Part 10

To accelerate the drying of the wood in the sand, the sand can be pre-warmed by placing it in a fireplace. The heat absorbed by the sand is then transferred to the wood, allowing it to dry faster. With this method, wood can be dried during any time of year and in any kind of weather.

Part 11

Of all these aforementioned good ways to dry timber, removing the tannin remains the best. Use the following equipment for this purpose:

1) *a* on Plate XVII, Fig. 1 is a copper kettle within walls, whose fireplace can be equipped with a grate to save fuel. On top of this kettle

2) there is a cap b^{56}, also made of copper or also strong sheet metal, which is placed on top of the lid c for the kettle a. The details of this cap are as follows: Depending on the size of the lower diameter of the cap, a hole is cut in the lid, over which the cap is positioned. A leather disk is placed under the 2-inch protruding rim d of the cap. With this device, the cap can be attached to the lid around its periphery with screws, resulting in an airtight seal. So that the lid remains in position and is not lifted off by the steam, a wall has to be constructed around it, as shown by e in Fig. 1.

3) A pipe f^{57} comes out of the cap b, which conducts the steam into the tannin-removal box (whose construction will be described below). This pipe can be 4 to 5 inches in diameter and must be securely soldered onto the cap.

4) So that the steam has a way out when the pressure becomes too great, a pipe is attached on top of the cap, with a lead ball on top of the pipe. This lead ball has a hole with a vertical wire h going through the middle. Steam can escape as the ball moves up and down. To ensure that this ball makes an airtight seal with the opening of the pipe g, the ball is first wrapped in fine oakum and then covered with leather, so that it forms a tight seal everywhere.

5) To be able to replace the water that evaporates from the kettle a, a pipe i is stuck through the lid[58]. This pipe must have a rim where it goes through the lid, in the same way that the cap has a rim, so that a leather strip can be placed below it and the pipe can be attached with screws (to make an airtight seal). The end of this pipe projects below the lower surface of the lid a distance that is one-third the height of the kettle. By the way, a cork is placed on the opening of the pipe to make the unit airtight.

To determine if it is necessary to refill the water in the kettle, take a stick and insert it through the pipe i into the kettle. From the wet part of the stick you will easily be able to tell if there is still enough water in the kettle. The other parts of the kettle and the firebox can easily be seen in the drawing.

Part 12

We now come to the second part of this machine, which consists of the tannin-removal box. The construction of this box is as follows. The box is made of strong columns, as shown in Fig. 3 and 4,[59] whose dimensions can be determined from the scale bar. In addition, however, I have to mention the following components:

1) The planks e in Fig. 2 and 3 must be joined by means of tongues and grooves on the sides and on the bottom as tightly as possible to prevent the steam from escaping.
2) On both gabled ends, the box has strong doors, which must be fitted closely between the planks. In addition, the joints must be painted with clay mixed with bovine blood.
3) This box must hang on a slant so that the tannin driven out of the wood by the steam can run out through a 1/8-inch wide by 2 inch high slot in the door.

---

[56] Translator's note: I don't see the designation b in the drawing.
[57] Translator's note: In the drawing, it looks like the pipe is k not f.
[58] Translator's note: In the drawing, it looks like the pipe i is attached to the kettle a, not the lid c.
[59] Translator's note: I don't see Fig. 3 and Fig. 4 on Plate XVII.

4) The lower end of the bottom of the box is set in a clay puddle[60] and the raised end is covered with the same material to prevent steam from escaping.
5) In addition, all side joints on the entire box must be puttied and painted thoroughly. All other components can be seen in the drawing.

Part 13

Method for removing tannin with this machine:

1) Put the wood from which tannin is to be removed in the box. The wood can be cut planks, boards, or other wood. The planks and boards should always stand on edge and the entire box should be tightly packed with wood, if possible, with the planks and boards standing close together. In addition, it should be mentioned that, after the box is filled with wood, the joints on the door must be carefully sealed. Now add
2) water to the kettle a through pipe i so that the level is one to two inches over the bottom opening of the pipe i. After doing this, make a fire below the kettle and then the tannin removal process can begin. Every hour or so, add enough water through pipe i to maintain the aforementioned water level. After 5 or 6 hours of heating, you will notice a very black and thick lye-like liquid running out of the small slot in the door. Now continue heating continuously until the liquid looks completely clear, without any color, just like water. This takes approximately 40 to 50 hours for planks and boards. If you want to make sure that tannin is being removed from the wood in the box, open the door of the box about 12 hours after starting the heating process. You should be able to see the tannin being excreted from the pores in the wood. But don't open the box again; otherwise the tannin removal process could be compromised. The fuel can be wood or peat, although peat (or brown coal bricks) is preferred because it produces a more even heat.

---

[60] Technical annotation: The strict translation of "clay puddle" can be interpreted as a clay lining or cover on the inside and outside of the box to seal it and prevent leaking. In this sense it would act as an early refractory material as well.

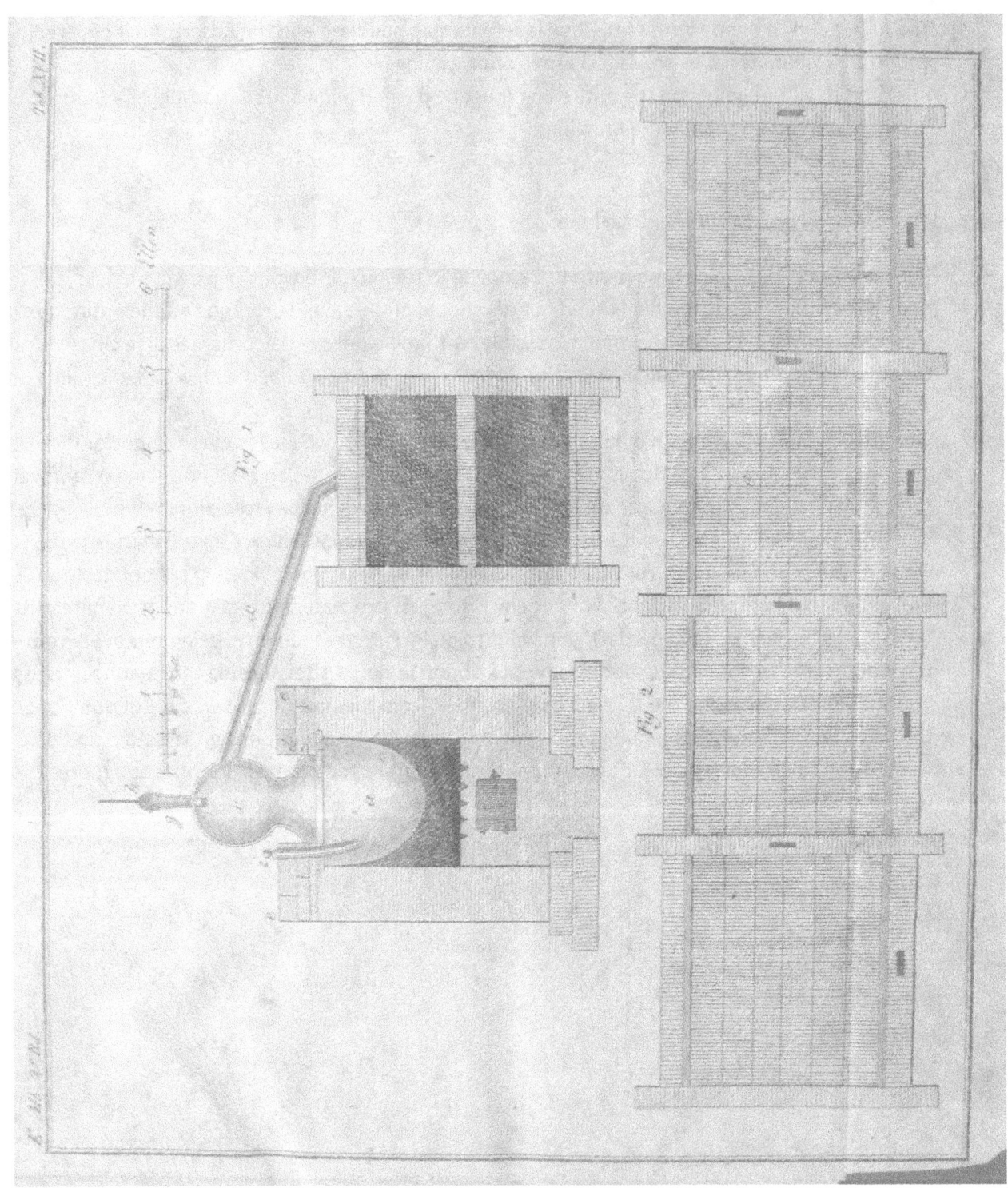

Plate XVII: Figures 1 and 2

Translator:

Karin I. Knisely is Lab Director for Core Course Biology at Bucknell University. One of her areas of interest is German-to-English translation, specializing in scientific and technical texts. She is an American Translators Association-certified translator. She obtained her M.S. degree in zoology at the University of New Hampshire. She is the author of *A Student Handbook for Writing in Biology* (Sinauer Associates, Inc./W.H. Freeman and Co., 2013), now in its Fourth Edition.
Contact: kknisely@bucknell.edu

Technical Annotator:

Thomas P. Rich is professor emeritus of mechanical engineering at Bucknell University. He served as the dean of engineering at Bucknell for 11 years, and held the Rooke Chair in the Historical and Social Context of Engineering. He obtained his undergraduate degree in mechanical engineering at Carnegie Mellon University, and he received his Ph.D. in mechanical engineering from Lehigh University. He is the co-author of a book published in 2014 by the Union County Historical Society, *Water-powered Gristmills of Union County, Pennsylvania* (www.unioncountyhistoricalsociety.org).
Contact: rich@bucknell.edu

www.ingramcontent.com/pod-product-compliance
Lightning Source LLC
Chambersburg PA
CBHW081733170526
45167CB00009B/3802